SUN TZU ON THE ART OF WAR

He will win who knows when to fight - and when not to fight. This is the classic translation of the Chinese military masterpiece, and it remains the best as it preserves the character and nuances of the Chinese original. The inspiration of Mao Tse Tung and countless generations of military leaders, it was written in antiquity and consists of thirteen chapters that reflect the mind of a born strategist and practical soldier whose maxims, full of acuteness and common sense, relate as much to the present day as they do to the military conditions of the time when they were written. Preceded by critical notes on Sun Tzu and the history of his work and including key Chinese characters, chapters are devoted to laying plans; waging war; attack by strategem, tactical dispositions, energy, weak points and strong; manoeuvering, variation of tactics, the army on the march; terrain, the nine situations, attack by fire and the use of spies, followed by an extensive index. As useful in the pursuit of success in modern business as it was in ancient warfare, this volume also relates to all aspects of personal and everyday life in which you must either be a winner or a loser.

LIONEL GILES was on the staff of the Department of Oriental Printed Books and Manuscripts, the British Museum.

The Kegan Paul China Library

WANDERINGS IN CHINA
Robert Fortune

VILLAGE LIFE IN CHINA
Arthur H. Smith

SOCIAL LIFE OF THE CHINESE
Justus Doolittle

RULERS OF CHINA
A. C. Moule

BEYOND THE GOLDEN LOTUS
Ko Lien Hua Ying

SUN TZU ON THE ART OF WAR

THE OLDEST MILITARY TREATISE IN THE WORLD

SUN TZU

Translated from the Chinese with
introduction and critical notes by Lionel Giles.

Routledge
Taylor & Francis Group

LONDON AND NEW YORK

First published in 2002 by
Kegan Paul Limited

Published 2014 by Routledge

2 Park Square, Milton Park, Abingdon, Oxfordshire OX14 4RN

711 Third Avenue, New York, NY 10017

First issued in paperback 2014

Routledge is an imprint of the Taylor & Francis Group, an informa business

© Taylor & Francis, 2002

British Library Cataloguing in Publication Data
Sun Tzu on the art of war : the oldest military treatise in
the world - (Kegan Paul life strategies)
1.Military art and science - China - Early works to 1800
I.Giles, Lionel
355'.02'0951
ISBN 978-0-710-30738-5 (hbk)

ISBN 978-1-138-86360-6 (pbk)

Library of Congress Cataloging-in-Publication Data
Applied for.

To my brother

Captain Valentine Giles, R.E.

in the hope that

a work 2400 years old

may yet contain lessons worth consideration

by the soldier of to-day

this translation

is affectionately dedicated

CONTENTS

		Page
PREFACE		vii
INTRODUCTION		
	Sun Wu and his Book	xi
	The Text of Sun Tzŭ	xxx
	The Commentators.	xxxiv
	Appreciations of Sun Tzŭ	xlii
	Apologies for war	xliii
	Bibliography	1
Chap.	I. Laying Plans	1
„	II. Waging War.	9
„	III. Attack by Stratagem.	17
„	IV. Tactical Dispositions.	26
„	V. Energy	33
„	VI. Weak Points and Strong	42
„	VII. Manœuvring.	55
„	VIII. Variation of Tactics.	71
„	IX. The Army on the March	80
„	X. Terrain	100
„	XI. The Nine Situations.	114
„	XII. The Attack by Fire	150
„	XIII. The Use of Spies.	160
CHINESE CONCORDANCE		176
INDEX		192

PREFACE

The seventh volume of "Mémoires concernant l'histoire, les sciences, les arts, les mœurs, les usages, &c., des Chinois"[1] is devoted to the Art of War, and contains, amongst other treatises, "Les Treize Articles de Sun-tse," translated from the Chinese by a Jesuit Father, Joseph Amiot. Père Amiot appears to have enjoyed no small reputation as a sinologue in his day, and the field of his labours was certainly extensive. But his so-called translation of Sun Tzŭ, if placed side by side with the original, is seen at once to be little better than an imposture. It contains a great deal that Sun Tzŭ did not write, and very little indeed of what he did. Here is a fair specimen, taken from the opening sentences of chapter 5 : —

De l'habileté dans le gouvernement des Troupes. Sun-tse dit: Ayez les noms de tous les Officiers tant généraux que subalternes; inscrivez-les dans un catalogue à part, avec la note des talents & de la capacité de chacun d'eux, afin de pouvoir les employer avec avantage lorsque l'occasion en sera venue. Faites en sorte que tous ceux que vous devez commander soient persuadés que votre principale attention est de les préserver de tout dommage. Les troupes que vous ferez avancer contre l'ennemi doivent être comme des pierres que vous lanceriez contre des œufs. De vous à l'ennemi il ne doit y avoir d'autre différence que celle du fort au foible, du vuide au plein. Attaquez à découvert, mais soyez vainqueur en secret. Voilà en peu de mots en quoi consiste l'habileté & toute la perfection même du gouvernement des troupes.

Throughout the nineteenth century, which saw a wonderful development in the study of Chinese literature, no translator ventured to tackle Sun Tzŭ, although his work was known to be highly valued in China as by far the

---

[1] Published at Paris in 1782.

oldest and best compendium of military science. It was
not until the year 1905 that the first English translation,
by Capt. E. F. Calthrop, R.F.A., appeared at Tokyo
under the title "Sonshi" (the Japanese form of Sun Tzŭ).[1]
Unfortunately, it was evident that the translator's knowledge of Chinese was far too scanty to fit him to grapple
with the manifold difficulties of Sun Tzŭ. He himself
plainly acknowledges that without the aid of two Japanese
gentlemen "the accompanying translation would have been
impossible." We can only wonder, then, that with their
help it should have been so excessively bad. It is not
merely a question of downright blunders, from which none
can hope to be wholly exempt. Omissions were frequent;
hard passages were wilfully distorted or slurred over. Such
offences are less pardonable. They would not be tolerated
in any edition of a Greek or Latin classic, and a similar
standard of honesty ought to be insisted upon in translations from Chinese.

From blemishes of this nature, at least, I believe that
the present translation is free. It was not undertaken
out of any inflated estimate of my own powers; but I
could not help feeling that Sun Tzŭ deserved a better
fate than had befallen him, and I knew that, at any rate,
I could hardly fail to improve on the work of my predecessors. Towards the end of 1908, a new and revised edition
of Capt. Calthrop's translation was published in London,
this time, however, without any allusion to his Japanese
collaborators. My first three chapters were then already
in the printer's hands, so that the criticisms of Capt.
Calthrop therein contained must be understood as referring to his earlier edition. In the subsequent chapters I
have of course transferred my attention to the second
edition. This is on the whole an improvement on the
other, though there still remains much that cannot pass

---

[1] A rather distressing Japanese flavour pervades the work throughout. Thus, King
Ho Lu masquerades as "Katsuryo," Wu and Yüeh become "Go" and "Etsu," etc. etc.

muster. Some of the grosser blunders have been rectified and lacunae filled up, but on the other hand a certain number of new mistakes appear. The very first sentence of the introduction is startlingly inaccurate; and later on, while mention is made of "an army of Japanese commentators" on Sun Tzŭ (who are these, by the way?), not a word is vouchsafed about the Chinese commentators, who nevertheless, I venture to assert, form a much more numerous and infinitely more important "army."

A few special features of the present volume may now be noticed. In the first place, the text has been cut up into numbered paragraphs, both in order to facilitate cross-reference and for the convenience of students generally. The division follows broadly that of Sun Hsing-yen's edition; but I have sometimes found it desirable to join two or more of his paragraphs into one. In quoting from other works, Chinese writers seldom give more than the bare title by way of reference, and the task of research is apt to be seriously hampered in consequence. With a view to obviating this difficulty so far as Sun Tzŭ is concerned, I have also appended a complete concordance of Chinese characters, following in this the admirable example of Legge, though an alphabetical arrangement has been preferred to the distribution under radicals which he adopted. Another feature borrowed from "The Chinese Classics" is the printing of text, translation and notes on the same page; the notes, however, are inserted, according to the Chinese method, immediately after the passages to which they refer. From the mass of native commentary my aim has been to extract the cream only, adding the Chinese text here and there when it seemed to present points of literary interest. Though constituting in itself an important branch of Chinese literature, very little commentary of this kind has hitherto been made directly accessible by translation. [1]

---

[1] A notable exception is to be found in Biot's edition of the *Chou Li*.

I may say in conclusion that, owing to the printing off of my sheets as they were completed, the work has not had the benefit of a final revision. On a review of the whole, without modifying the substance of my criticisms, I might have been inclined in a few instances to temper their asperity. Having chosen to wield a bludgeon, however, I shall not cry out if in return I am visited with more than a rap over the knuckles. Indeed, I have been at some pains to put a sword into the hands of future opponents by scrupulously giving either text or reference for every passage translated. A scathing review, even from the pen of the Shanghai critic who despises "mere translations," would not, I must confess, be altogether unwelcome. For, after all, the worst fate I shall have to dread is that which befel the ingenious paradoxes of George in *The Vicar of Wakefield.*

INTRODUCTION

SUN WU AND HIS BOOK.

Ssŭ-ma Ch'ien gives the following biography of Sun Tzŭ: [1] —

孫 子 武 Sun Tzŭ Wu was a native of the Ch'i State. His *Art of War* brought him to the notice of 闔 廬 Ho Lu,[2] King of 吳 Wu. Ho Lu said to him: I have carefully perused your 13 chapters. May I submit your theory of managing soldiers to a slight test? — Sun Tzŭ replied: You may. — Ho Lu asked: May the test be applied to women? — The answer was again in the affirmative, so arrangements were made to bring 180 ladies out of the Palace. Sun Tzŭ divided them into two companies, and placed one of the King's favourite concubines at the head of each. He then bade them all take spears in their hands, and addressed them thus: I presume you know the difference between front and back, right hand and left hand? — The girls replied: Yes. — Sun Tzŭ went on: When I say "Eyes front," you must look straight ahead. When I say "Left turn," you must face towards your left hand. When I say "Right turn," you must face towards your right hand. When I say "About turn," you must face right round towards the back. — Again the girls assented. The words of command having been thus ex- plained, he set up the halberds and battle-axes in order to begin the drill. Then, to the sound of drums, he gave the order "Right turn." But the girls only burst out laughing. Sun Tzŭ said: If words of command are not clear and distinct, if orders are not thoroughly understood, then the general is to blame. — So he started drilling them again, and this time gave the order "Left turn," whereupon the girls once more burst into fits of laughter. Sun Tzŭ said: If words of command are not clear and distinct, if orders are not thoroughly understood, the general is to blame. But if his orders *are* clear, and the soldiers nevertheless disobey, then it is the fault of their officers. — So saying, he ordered the leaders of the two companies to be beheaded. Now the King of Wu was watching the

---

[1] *Shih Chi*, ch. 65.

[2] Also written 闔 閭 Ho Lü. He reigned from 514 to 496 B.C.

scene from the top of a raised pavilion; and when he saw that his favourite concubines were about to be executed, he was greatly alarmed
and hurriedly sent down the following message: We are now quite satisfied as to our general's ability to handle troops. If We are bereft of these
two concubines, our meat and drink will lose their savour. It is our
wish that they shall not be beheaded. — Sun Tzŭ replied: Having once
received His Majesty's commission to be general of his forces, there are
certain commands of His Majesty which, acting in that capacity, I am
unable to accept. — Accordingly, he had the two leaders beheaded, and
straightway installed the pair next in order as leaders in their place.
When this had been done, the drum was sounded for the drill once more;
and the girls went through all the evolutions, turning to the right or to
the left, marching ahead or wheeling back, kneeling or standing, with
perfect accuracy and precision, not venturing to utter a sound. Then
Sun Tzŭ sent a messenger to the King saying: Your soldiers, Sire, are
now properly drilled and disciplined, and ready for Your Majesty's in-
spection. They can be put to any use that their sovereign may desire;
bid them go through fire and water, and they will not disobey. — But
the King replied: Let our general cease drilling and return to camp. As
for us, We have no wish to come down and inspect the troops. — Thereupon Sun Tzŭ said: The King is only fond of words, and cannot translate them into deeds. — After that, Ho Lu saw that Sun Tzŭ was one
who knew how to handle an army, and finally appointed him general.
In the West, he defeated the Ch'u State and forced his way into Ying,
the capital; to the north, he put fear into the States of Ch'i and Chin,
and spread his fame abroad amongst the feudal princes. And Sun Tzŭ
shared in the might of the King.

About Sun Tzŭ himself this is all that Ssŭ-ma Ch'ien
has to tell us in this chapter. But he proceeds to give
a biography of his descendant, 孫臏 Sun Pin, born about
a hundred years after his famous ancestor's death, and
also the outstanding military genius of his time. The
historian speaks of him too as Sun Tzŭ, and in his preface
we read: 孫子臏腳而論兵法 "Sun Tzŭ had his feet
cut off and yet continued to discuss the art of war." [1]
It seems likely, then, that "Pin" was a nickname bestowed
on him after his mutilation, unless indeed the story was
invented in order to account for the name. The crowning
incident of his career, the crushing defeat of his treacherous
rival P'ang Chüan, will be found briefly related on p. 40.

[1] *Shih Chi*, ch. 130, f. 6 r°.

To return to the elder Sun Tzŭ. He is mentioned in two other passages of the *Shih Chi:* —

In the third year of his reign [512 B.C.] Ho Lu, King of Wu, took the field with 子胥 Tzŭ-hsü [i.e. 伍員 Wu Yüan] and 伯嚭 Po P'ei, and attacked Ch'u. He captured the town of 舒 Shu and slew the two prince's sons who had formerly been generals of Wu. He was then meditating a descent on 郢 Ying [the capital]; but the general Sun Wu said: "The army is exhausted. [1] It is not yet possible. We must wait".... [2] [After further successful fighting,] "in the ninth year [506 B.C.], King Ho Lu of Wu addressed Wu Tzŭ-hsü and Sun Wu, saying: "Formerly, you declared that it was not yet possible for us to enter Ying. Is the time ripe now?" The two men replied: "Ch'u's general, 子常 Tzŭ-ch'ang, [3] is grasping and covetous, and the princes of 唐 T'ang and 蔡 Ts'ai both have a grudge against him. If Your Majesty has resolved to make a grand attack, you must win over T'ang and Ts'ai, and then you may succeed." Ho Lu followed this advice, [beat Ch'u in five pitched battles and marched into Ying]. [4]

This is the latest date at which anything is recorded of Sun Wu. He does not appear to have survived his patron, who died from the effects of a wound in 496.

In the chapter entitled 律書 (the earlier portion of which M. Chavannes believes to be a fragment of a treatise on Military Weapons), there occurs this passage: [5]

From this time onward, a number of famous soldiers arose, one after the other: 咎犯 Kao-fan, [6] who was employed by the Chin State; Wang-tzŭ, [7] in the service of Ch'i; and Sun Wu, in the service of Wu. These men developed and threw light upon the principles of war (申 明軍約).

---

[1] I note that M. Chavannes translates 民勞 "le peuple est épuisé." But in Sun Tzŭ's own book (see especially VII §§ 24—26) the ordinary meaning of 民 is "army," and this, I think, is more suitable here.

[2] These words are given also in Wu Tzŭ-hsü's biography, ch. 66, fol. 3 *r*°.

[3] The appellation of 囊瓦 Nang Wa.

[4] *Shih Chi*, ch. 31, fol. 6 *r*°.

[5] *Ibid.* ch. 25, fol. 1 *r*°.

[6] The appellation of 狐偃 Hu Yen, mentioned in ch. 39 under the year 637.

[7] 王子城父 Wang-tzŭ Ch'êng-fu, ch. 32, year 607.

It is obvious that Ssŭ-ma Ch'ien at least had no doubt about the reality of Sun Wu as an historical personage; and with one exception, to be noticed presently, he is by far the most important authority on the period in question. It will not be necessary, therefore, to say much of such a work as the 吳越春秋 *Wu Yüeh Ch'un Ch'iu*, which is supposed to have been written by 趙曄 Chao Yeh of the 1st century A.D. The attribution is somewhat doubtful; but even if it were otherwise, his account would be of little value, based as it is on the *Shih Chi* and expanded with romantic details. The story of Sun Tzŭ will be found, for what it is worth, in chapter 2. The only new points in it worth noting are: 1) Sun Tzŭ was first recommended to Ho Lu by Wu Tzŭ-hsü. 2) He is called a native of Wu.[1] 3) He had previously lived a retired life, and his contemporaries were unaware of his ability.[2]

The following passage occurs in 淮南子 Huai-nan Tzŭ: "When sovereign and ministers show perversity of mind, it is impossible even for a Sun Tzŭ to encounter the foe."[3] Assuming that this work is genuine (and hitherto no doubt has been cast upon it), we have here the earliest direct reference to Sun Tzŭ, for Huai-nan Tzŭ died in 122 B.C., many years before the *Shih Chi* was given to the world.

劉向 Liu Hsiang (B.C. 80–9) in his 新序 says: "The reason why Sun Wu at the head of 30,000 men beat

---

[1] The mistake is natural enough. Native critics refer to the 越絕書, a work of the Han dynasty, which says (ch. 2, fol. 3 *v°* of my edition): 巫門外大冢吳王客齊孫武冢也去縣十里善為兵法 "Ten *li* outside the *Wu* gate [of the city of 吳 Wu, now Soochow in Kiangsu] there is a great mound, raised to commemorate the entertainment of Sun Wu of Ch'i, who excelled in the art of war, by the King of Wu."

[2] 孫子者吳人也善為兵法辟幽居世人莫知其能.

[3] 君臣乖心則孫子不能以應敵.

Ch'u with 200,000 is that the latter were undisciplined."[1]

鄧名世 Têng Ming-shih in his 姓氏辨證書 (completed in 1134) informs us that the surname 孫 was bestowed on Sun Wu's grandfather by 景公 Duke Ching of Ch'i [547–490 B.C.]. Sun Wu's father Sun 馮 P'ing, rose to be a Minister of State in Ch'i, and Sun Wu himself, whose style was 長卿 Ch'ang-ch'ing, fled to Wu on account of the rebellion which was being fomented by the kindred of 田鮑 T'ien Pao. He had three sons, of whom the second, named 明 Ming, was the father of Sun Pin. According to this account, then, Pin was the grandson of Wu,[2] which, considering that Sun Pin's victory over 魏 Wei was gained in 341 B.C., may be dismissed as chronologically impossible. Whence these data were obtained by Têng Ming-shih I do not know, but of course no reliance whatever can be placed in them.

An interesting document which has survived from the close of the Han period is the short preface written by the great 曹操 Ts'ao Ts'ao, or 魏武帝 Wei Wu Ti, for his edition of Sun Tzŭ. I shall give it in full: —

I have heard that the ancients used bows and arrows to their advantage.[3] The *Lun Yü* says: "There must be a sufficiency of military strength."[4] The *Shu Ching* mentions "the army" among the "eight objects of government."[5] The *I Ching* says: "師 'army' indicates firmness and justice; the experienced leader will have good fortune."[6]

---

[1] 孫武以三萬破楚二十萬者楚無法故也·

[2] The *Shih Chi*, on the other hand, says: 臏亦孫武之後世子孫也. I may remark in passing that the name 武 for one who was a great warrior is just as suspicious as 臏 for a man who had his feet cut off.

[3] An allusion to 易經, 繫辭, II. 2: 弦木爲弧剡木爲矢弧矢之利以威天下 "They attached strings to wood to make bows, and sharpened wood to make arrows. The use of bows and arrows is to keep the Empire in awe."

[4] 論語 XII. 7.   [5] 書經 V. iv. 7.

[6] 易經, 7th diagram (師).

The *Shih Ching* says: "The King rose majestic in his wrath, and he marshalled his troops." [1] The Yellow Emperor, T'ang the Completer and Wu Wang all used spears and battle-axes in order to succour their generation. The *Ssŭ-ma Fa* says: "If one man slay another of set purpose, he himself may rightfully be slain." [2] He who relies solely on warlike measures shall be exterminated; he who relies solely on peaceful measures shall perish. Instances of this are Fu Ch'ai [3] on the one hand and Yen Wang on the other. [4] In military matters, the Sage's rule is normally to keep the peace, and to move his forces only when occasion requires. He will not use armed force unless driven to it by necessity. [5]

Many books have I read on the subject of war and fighting; but the work composed by Sun Wu is the profoundest of them all. [Sun Tzŭ was a native of the Ch'i state, his personal name was Wu. He wrote the *Art of War* in 13 chapters for Ho Lü, King of Wu. Its principles were tested on women, and he was subsequently made a general. He led an army westwards, crushed the Ch'u State and entered Ying the capital. In the north, he kept Ch'i and Chin in awe. A hundred years and more after his time, Sun Pin lived. He was a descendant of Wu]. [6] In his treatment of deliberation and planning, the importance of rapidity in taking the field, [7] clearness of conception, and depth of design, Sun

---

[1] 詩經 III. 1. vii. 5.

[2] 司馬法 ch. 1 (仁本) *ad init.* The text of the passage in the 圖書 *T'u Shu* (戎政典, ch. 85) is: 是故殺人安人殺之可也.

[3] The son and successor of Ho Lu. He was finally defeated and overthrown by 勾踐 Kou Chien, King of Yüeh, in 473 B.C. See *post.*

[4] King Yen of 徐 Hsü, a fabulous being, of whom Sun Hsing-yen says in his preface: 仁而敗 "His humanity brought him to destruction." See *Shih Chi*, ch. 5, f. 1 *v*°, and M. Chavannes' note, *Mémoires Historiques*, tom. II, p. 8.

[5] *T'u Shu, ibid.* ch. 90: 樑聞上古有弧矢之利論語曰足兵尚書八政曰師易曰師貞丈人吉詩曰王赫斯怒爰征其旅黃帝湯武咸用干戚以濟世也司馬法曰人故殺人殺之可也恃武者滅恃文者亡夫差偃王是也聖人之用兵戢而時動不得已而用之.

[6] The passage I have put in brackets is omitted in the *T'u Shu*, and may be an interpolation. It was known, however, to 張守節 Chang Shou-chieh of the T'ang dynasty, and appears in the *T'ai P'ing Yü Lan.*

[7] Ts'ao Kung seems to be thinking of the first part of chap. II, perhaps especially of § 8.

Tzŭ stands beyond the reach of carping criticism. My contemporaries, however, have failed to grasp the full meaning of his instructions, and while putting into practice the smaller details in which his work abounds, they have overlooked its essential purport. That is the motive which has led me to outline a rough explanation of the whole. [1]

One thing to be noticed in the above is the explicit statement that the 13 chapters were specially composed for King Ho Lu. This is supported by the internal evidence of I. § 15, in which it seems clear that some ruler is addressed.

In the bibliographical section of the *Han Shu*, [2] there is an entry which has given rise to much discussion: 吳孫子八十二篇圖九卷 "The works of Sun Tzŭ of Wu in 82 *p'ien* (or chapters), with diagrams in 9 *chüan*." It is evident that this cannot be merely the 13 chapters known to Ssŭ-ma Ch'ien, or those we possess to-day. Chang Shou-chieh in his 史記正義 refers to an edition of Sun Tzŭ's 兵法 of which the "13 chapters" formed the first *chüan*, adding that there were two other *chüan* besides. [3] This has brought forth a theory, that the bulk of these 82 chapters consisted of other writings of Sun Tzŭ — we should call them apocryphal — similar to the 問答 *Wên Ta*, of which a specimen dealing with the Nine Situations [4] is preserved in the 通典 *Tung Tien*, and another in Ho Shih's commentary. It is suggested

---

[1] 吾觀兵書戰策多矣孫武所著深矣孫子者齊人也名武為吳王闔閭作兵法一十三篇試之婦人卒以為將西破強楚入郢北威齊晉後百歲餘有孫臏是武之後也審計重舉明畫深圖不可相誣而但世人未之深亮訓說況文煩富行於世者失其旨要故撰為略解焉.

[2] 漢書藝文志、兵權謀.

[3] The 宋藝文志 mentions two editions of Sun Tzŭ in 3 *chüan*, namely 孫武孫子 and 朱服校定孫子.

[4] See chap. XI.

that before his interview with Ho Lu, Sun Tzŭ had only written the 13 chapters, but afterwards composed a sort of exegesis in the form of question and answer between himself and the King. 畢 以 珣 Pi I-hsün, author of the 孫子敍錄 Sun Tzŭ Hsü Lu, backs this up with a quotation from the Wu Yüeh Ch'un Ch'iu: "The King of Wu summoned Sun Tzŭ, and asked him questions about the art of war. Each time he set forth a chapter of his work, the King could not find words enough to praise him." [1] As he points out, if the whole work was expounded on the same scale as in the above-mentioned fragments, the total number of chapters could not fail to be considerable. [2] Then the numerous other treatises attributed to Sun Tzŭ [3] might also be included. The fact that the Han Chih mentions no work of Sun Tzŭ except the 82 p'ien, whereas the Sui and T'ang bibliographies give the titles of others in addition to the "13 chapters," is good proof, Pi I-hsün thinks, that all of these were contained in the 82 p'ien. Without pinning our faith to the accuracy of details supplied by the Wu Yüeh Ch'un Ch'iu, or admitting the genuineness of any of the treatises cited by Pi I-hsün, we may see in this theory a probable solution of the mystery. Between Ssŭ-ma Ch'ien and Pan Ku there was plenty of time for a luxuriant crop of forgeries to have grown up under the magic name of Sun Tzŭ, and the 82 p'ien may very well represent a collected edition of these lumped together with the original work.

_____

[1] 吳王召孫子問以兵法每陳一篇王不知口之稱善·

[2] 按此皆釋九地篇義辭意甚詳故其篇帙不能不多也·

[3] Such as the 八陣圖, quoted in 鄭玄 Chêng Hsüan's commentary on the Chou Li, the 戰鬭大甲兵法 and 兵法雜占, mentioned in the 隋志 Sui Chih, and the 三十二壘經, in the Hsin T'ang Chih.

It is also possible, though less likely, that some of them existed in the time of the earlier historian and were purposely ignored by him. [1]

Tu Mu, after Ts'ao Kung the most important commentator on Sun Tzŭ, composed the preface to his edition [2] about the middle of the ninth century. After a somewhat lengthy defence of the military art, [3] he comes at last to Sun Tzŭ himself, and makes one or two very startling assertions: — "The writings of Sun Wu," he says, "originally comprised several hundred thousand words, but Ts'ao Ts'ao, the Emperor Wu Wei, pruned away all redundancies and wrote out the essence of the whole, so as to form a single book in 13 chapters." [4] He goes on to remark that Ts'ao Ts'ao's commentary on Sun Tzŭ leaves a certain proportion of difficulties unexplained. This, in Tu Mu's opinion, does not necessarily imply that he was unable to furnish a complete commentary. [5] According to the *Wei Chih*, Ts'ao himself wrote a book on war in something over 100,000 words, known as the 新書. It appears to have been of such exceptional merit that he suspects Ts'ao to have used for it the surplus material which he had found in Sun Tzŭ. He concludes, however, by saying: "The *Hsin Shu* is now lost, so that the truth cannot be known for certain." [6]

Tu Mu's conjecture seems to be based on a passage

---

[1] On the other hand, it is noteworthy that 吳子 *Wu Tzŭ*, which is now in 6 chapters, has 48 assigned to it in the *Han Chih*. Likewise, the 中庸 *Chung Yung* is credited with 49 chapters, though now in one only. In the case of such very short works, one is tempted to think that 篇 might simply mean "leaves."

[2] See *T'u Shu*, 經籍典, ch. 442, 彙考 2.

[3] An extract will be found on p. xlv.

[4] 武所著書凡數十萬言曹魏武帝削其繁剩筆其精切凡十三篇成爲一編.

[5] 其所爲注解十不釋一此蓋非曹不能盡注解也.

[6] 予尋魏志見曹自作兵書十餘萬言諸將

in the 漢官解詁 "Wei Wu Ti strung together Sun Wu's Art of War,"[1] which in turn may have resulted from a misunderstanding of the final words of Ts'ao Kung's preface: 故撰爲略解焉. This, as Sun Hsing-yen points out,[2] is only a modest way of saying that he made an explanatory paraphrase,[3] or in other words, wrote a commentary on it. On the whole, the theory has met with very little acceptance. Thus, the 四庫全書 says:[4] "The mention of the 13 chapters in the *Shih Chi* shows that they were in existence before the *Han Chih*, and that later accretions are not to be considered part of the original work. Tu Mu's assertion can certainly not be taken as proof."[5]

There is every reason to suppose, then, that the 13 chapters existed in the time of Ssŭ-ma Ch'ien practically as we have them now. That the work was then well known he tells us in so many words: "Sun Tzŭ's 13 Chapters and Wu Ch'i's Art of War are the two books that people commonly refer to on the subject of military matters. Both of then are widely distributed, so I will not discuss them here."[6] But as we go further back, serious difficulties begin to arise. The salient fact which has to be faced is that the *Tso Chuan*, the great contemporary record, makes no mention whatever of Sun

---

征戰皆以新書從事從令者克捷違教者貶敗
意曹自於新書中馳騁其說自成一家事業不
欲隨孫武後盡解其書不然者曹其不能耶今
新書已亡不可復知.

[1] 魏氏瑣連孫武之法.　　　[2] See 孫子兵法序.

[3] 謙言解其梗略.　　　[4] Ch. 99, fol. 5 r°.

[5] 然史記稱十三篇在漢志之前不得以後
來附益者爲本書牧之言固未可以爲據也.

[6] *Shih Chi*, ch. 65 *ad fin:* 世俗所稱師旅皆道孫子十
三篇吳起兵法世多有故弗論.

Wu, either as a general or as a writer. It is natural, in view of this awkward circumstance, that many scholars should not only cast doubt on the story of Sun Wu as given in the *Shih Chi*, but even show themselves frankly sceptical as to the existence of the man at all. The most powerful presentment of this side of the case is to be found in the following disquisition by 葉水心 Yeh Shui-hsin:[1] —

It is stated in Ssŭ-ma Ch'ien's history that Sun Wu was a native of the Ch'i State, and employed by Wu; and that in the reign of Ho Lü he crushed Ch'u, entered Ying, and was a great general. But in Tso's Commentary no Sun Wu appears at all. It is true that Tso's Commentary need not contain absolutely everything that other histories contain. But Tso has not omitted to mention vulgar plebeians and hireling ruffians such as Ying K'ao-shu,[2] Ts'ao Kuei,[3] Chu Chih-wu[4] and Chuan Shê-chu.[5] In the case of Sun Wu, whose fame and achievements were so brilliant, the omission is much more glaring. Again, details are given, in their due order, about his contemporaries Wu Yüan and the Minister P'ei.[6] Is it credible that Sun Wu alone should have been passed over?[7]

In point of literary style, Sun Tzŭ's work belongs to the same school as *Kuan Tzŭ*,[8] the *Liu T'ao*,[9] and the *Yüeh Yü*,[10] and may have

---

[1] 葉適 Yeh Shih of the Sung dynasty [1151—1223]. See 文獻通考, ch. 221, ff. 7, 8.

[2] See *Tso Chuan*, 隱公, I. 3 *ad fin.* and XI. 3 *ad init.* He hardly deserves to be bracketed with assassins.

[3] See pp. 66, 128.

[4] See *Tso Chuan*, 僖公, XXX. 5.

[5] See p. 128. Chuan Chu is the abbreviated form of his name.

[6] *I. e.* Po P'ei. See *ante.*

[7] 遷載孫武齊人而用於吳在闔閭時破楚入郢爲大將按左氏無孫武他書所有左氏不必盡有然穎考叔曹劌燭之武鱄設諸之流微賤暴用事左氏未嘗遺而武功名章灼如此乃更關又同時伍員宰嚭一一銓次乃獨不及武邪.

[8] The nucleus of this work is probably genuine, though large additions have been made by later hands. Kuan Chung died in 645 B.C.

[9] See *infra*, p. 1.

[10] I do not know what work this is, unless it be the last chapter of the 國語. Why that chapter should be singled out, however, is not clear.

been the production of some private scholar living towards the end of the "Spring and Autumn" or the beginning of the "Warring States" period. [1] The story that his precepts were actually applied by the Wu State, is merely the outcome of big talk on the part of his followers. [2]

From the flourishing period of the Chou dynasty [3] down to the time of the "Spring and Autumn," all military commanders were statesmen as well, and the class of professional generals, for conducting external campaigns, did not then exist. It was not until the period of the "Six States" [4] that this custom changed. Now although Wu was an uncivilised State, is it conceivable that Tso should have left unrecorded the fact that Sun Wu was a great general and yet held no civil office? What we are told, therefore, about Jang-chü [5] and Sun Wu, is not authentic matter, but the reckless fabrication of theorising pundits. The story of Ho Lü's experiment on the women, in particular, is utterly preposterous and incredible. [6]

Yeh Shui-hsin represents Ssŭ-ma Ch'ien as having said that Sun Wu crushed Ch'u and entered Ying. This is not quite correct. No doubt the impression left on the reader's mind is that he at least shared in these exploits; but the actual subject of the verbs 破, 入, 威 and 顯 is certainly 闔廬, as is shown by the next words: 孫子與有力焉. [7] The fact may or may not be significant; but it is nowhere explicitly stated in the *Shih Chi* either that Sun Tzŭ was general on the occasion of

---

[1] About 480 B.C.

[2] 詳味孫子與管子六韜越語相出入春秋末戰國初山林處士所爲其言得用於吳者其徒夸大之說也.

[3] That is, I suppose, the age of Wu Wang and Chou Kung.

[4] In the 3rd century B.C.

[5] Ssŭ-ma Jang-chü, whose family name was 田 T'ien, lived in the latter half of the 6th century B.C., and is also believed to have written a work on war. See *Shih Chi*, ch. 64, and *infra*, p. 1.

[6] 自周之盛至春秋凡將兵者必與聞國政未有特將於外者六國時此制始改吳雖蠻夷而孫武爲大將乃不爲命卿而左氏無傳焉可乎故凡謂穰苴孫武者皆辯士妄相標指非事實其言闔閭試以婦人尤爲奇險不足信.

[7] See the end of the passage quoted from the *Shih Chi* on p. xii.

the taking of Ying, or that he even went there at all. Moreover, as we know that Wu Yüan and Po P‘ei both took part in the expedition, and also that its success was largely due to the dash and enterprise of 夫槩 Fu Kai, Ho Lu's younger brother, it is not easy to see how yet another general could have played a very prominent part in the same campaign.

陳振孫 Ch‘ên Chên-sun of the Sung dynasty has the note:[1] —

> Military writers look upon Sun Wu as the father of their art. But the fact that he does not appear in the *Tso Chuan*, although he is said to have served under Ho Lü King of Wu, makes it uncertain what period he really belonged to.[2]

He also says: —

> The works of Sun Wu and Wu Ch‘i may be of genuine antiquity.[3]

It is noticeable that both Yeh Shui-hsin and Ch‘ên Chên-sun, while rejecting the personality of Sun Wu as he figures in Ssŭ-ma Ch‘ien's history, are inclined to accept the date traditionally assigned to the work which passes under his name. The author of the *Hsü Lu* fails to appreciate this distinction, and consequently his bitter attack on Ch‘ên Chên-sun really misses its mark. He makes one or two points, however, which certainly tell in favour of the high antiquity of our "13 chapters." "Sun Tzŭ," he says, "must have lived in the age of Ching Wang [519–476], because he is frequently plagiarised in subsequent works of the Chou, Ch‘in and Han dynasties."[4]

---

[1] In the 書錄解題, a classified catalogue of his family library.
[2] See *Wên Hsien T‘ung K‘ao*, ch. 221, f. 9 r°: 世之言兵者祖孫武然孫武事吳闔閭而不見於左傳不知果何時人也.
[3] See *Hsü Lu*, f. 14 r°: 孫吳或是古書.
[4] 按孫子生於敬王之代故周秦兩漢諸書皆多襲用其文. Here is a list of the passages in Sun Tzŭ from which

The two most shameless offenders in this respect are Wu Ch'i and Huai-nan Tzŭ, both of them important historical personages in their day. The former lived only a century after the alleged date of Sun Tzŭ, and his death is known to have taken place in 381 B.C. It was to him, according to Liu Hsiang, that 曾申 Tsêng Shên delivered the *Tso Chuan*, which had been entrusted to him by its author.[1] Now the fact that quotations from the *Art of War*, acknowledged or otherwise, are to be found in so many authors of different epochs, establishes a very strong probability that there was some common source anterior to them all, — in other words, that Sun Tzŭ's treatise was already in existence towards the end of the 5th century B.C. Further proof of Sun Tzŭ's antiquity is furnished by the archaic or wholly obsolete meanings attaching to a number of the words he uses. A list of these, which might perhaps be extended, is given in the *Hsü Lu;* and though some of the interpretations are doubtful, the main argument is hardly affected thereby.[2] Again, it must not be forgotten that Yeh Shui-hsin, a scholar and critic of the first rank, deliberately pronounces the style of the 13 chapters to

---

either the substance or the actual words have been appropriated by early authors: VII. 9; IX. 17; I. 24 (戰國策). IX. 23; IX. 1, 3, 7; V. 1; III. 18; XI. 58; VII. 31; VII. 24; VII. 26; IX. 15; IX. 4 (*bis*) (吳子). III. 8; IV. 7 (尉繚子). VII. 19; V. 14; III 2 (鶡冠子). III. 8; XI. 2; I. 19; XI. 58; X. 10 & VI. 1 (史記. Two of the above are given as quotations). V. 13; IV. 2 (呂氏春秋). IX. 11, 12; XI. 30; I. 13; VII. 19 & IV. 7; VII. 32; VII. 25; IV. 20 & V. 23; IX. 43; V. 15; VII. 26; V. 4 & XI. 39; VIII. 11; VI. 4 (淮南子). V. 4 (太元經). II. 20; X. 14 (潛夫論).

[1] See Legge's Classics, vol. V, Prolegomena p. 27. Legge thinks that the *Tso Chuan* must have been written in the 5th century, but not before 424 B.C.

[2] The instances quoted are: — III. 14, 15: 同 is said to be equivalent to 冒; II. 15: 葸 = 其; VII. 28: 歸 = 息; XI. 60: 詳 = 佯; XI. 24: the use of 鬥 instead of 鬪 (the later form); XI. 64: 誅 = 治; IX. 3: 絕 = 越; III. 11: 周 and 隙 antithetically opposed in the sense of 無缺 and 有缺; XI. 56: 犯 = 動; XI. 31: 方 = 縛.

belong to the early part of the fifth century. Seeing that
he is actually engaged in an attempt to disprove the
existence of Sun Wu himself, we may be sure that he
would not have hesitated to assign the work to a later
date had he not honestly believed the contrary. And it
is precisely on such a point that the judgment of an
educated Chinaman will carry most weight. Other internal
evidence is not far to seek. Thus, in XIII. § 1, there is
an unmistakable allusion to the ancient system of land-
tenure which had already passed away by the time of
Mencius, who was anxious to see it revived in a modified
form. [1] The only warfare Sun Tzŭ knows is that carried
on between the various feudal princes (諸侯), in which
armoured chariots play a large part. Their use seems to
have entirely died out before the end of the Chou dynasty.
He speaks as a man of Wu, a state which ceased to
exist as early as 473 B. C. On this I shall touch presently.

But once refer the work to the 5th century or earlier,
and the chances of its being other than a *bonâ fide* production are sensibly diminished. The great age of forgeries
did not come until long after. That it should have been
forged in the period immediately following 473 is particularly unlikely, for no one, as a rule, hastens to identify
himself with a lost cause. As for Yeh Shui-hsin's theory,
that the author was a literary recluse, [2] that seems to me
quite untenable. If one thing is more apparent than another after reading the maxims of Sun Tzŭ, it is that their
essence has been distilled from a large store of personal
observation and experience. They reflect the mind not
only of a born strategist, gifted with a rare faculty of gene-
ralisation, but also of a practical soldier closely acquainted
with the military conditions of his time. To say nothing

---

[1] See *Mencius* III. 1. iii. 13—20.

[2] 山林處士 need not be pressed to mean an actual dweller in the
mountains. I think it simply denotes a person living a retired life and standing
aloof from public affairs.

of the fact that these sayings have been accepted and endorsed by all the greatest captains of Chinese history, they offer a combination of freshness and sincerity, acuteness and common sense, which quite excludes the idea that they were artificially concocted in the study. If we admit, then, that the 13 chapters were the genuine production of a military man living towards the end of the "Ch'un Ch'iu" period, are we not bound, in spite of the silence of the *Tso Chuan*, to accept Ssŭ-ma Ch'ien's account in its entirety? In view of his high repute as a sober historian, must we not hesitate to assume that the records he drew upon for Sun Wu's biography were false and untrustworthy? The answer, I fear, must be in the negative. There is still one grave, if not fatal, objection to the chronology involved in the story as told in the *Shih Chi*, which, so far as I am aware, nobody has yet pointed out. There are two passages in Sun Tzŭ in which he alludes to contemporary affairs. The first is in VI. § 21: —

> Though according to my estimate the soldiers of Yüeh exceed our own in number, that shall advantage them nothing in the matter of victory. I say then that victory can be achieved.

The other is in XI. § 30: —

> Asked if an army can be made to imitate the *shuai-jan*, I should answer, Yes. For the men of Wu and the men of Yüeh are enemies; yet if they are crossing a river in the same boat and are caught by a storm, they will come to each other's assistance just as the left hand helps the right.

These two paragraphs are extremely valuable as evidence of the date of composition. They assign the work to the period of the struggle between Wu and Yüeh. So much has been observed by Pi I-hsün. But what has hitherto escaped notice is that they also seriously impair the credibility of Ssŭ-ma Ch'ien's narrative. As we have seen above, the first positive date given in connection with Sun Wu is 512 B.C. He is then spoken of as a general, acting as confidential adviser to Ho Lu, so that his alleged introduction to that monarch had already taken place, and of course the 13 chapters must have been written earlier still. But at that time, and for several years after, down to the capture of Ying in 506, 楚 Ch'u, and not Yüeh, was the great hereditary enemy of Wu. The two states, Ch'u and Wu, had been constantly at war for over half a century,[1] whereas the first war between Wu and Yüeh was waged only in 510,[2] and even then was no more than a short interlude sandwiched in the midst of the fierce struggle with Ch'u. Now Ch'u is not mentioned in the 13 chapters at all. The natural inference is that they were written at a time when Yüeh had become the prime antagonist of Wu, that is, after Ch'u had suffered the great humiliation of 506. At this point, a table of dates may be found useful.

B.C.	
514	Accession of Ho Lu.
512	Ho Lu attacks Ch'u, but is dissuaded from entering 郢 Ying, the capital. *Shih Chi* mentions Sun Wu as general.
511	Another attack on Ch'u.
510	Wu makes a successful attack on Yüeh. This is the first war between the two states.
509 or 508	Ch'u invades Wu, but is signally defeated at 豫章 Yü-chang.
506	Ho Lu attacks Ch'u with the aid of T'ang and Ts'ai. Decisive battle of 柏舉 Po-chü, and capture of Ying. Last mention of Sun Wu in *Shih Chi*.
505	Yüeh makes a raid on Wu in the absence of its army. Wu is beaten by Ch'in and evacuates Ying.
504	Ho Lu sends 夫差 Fu Ch'ai to attack Ch'u.
497	勾踐 Kou Chien becomes King of Yüeh.
496	Wu attacks Yüeh, but is defeated by Kou Chien at 檇李 Tsui-li. Ho Lu is killed.

---

[1] When Wu first appears in the *Ch'un Ch'iu* in 584, it is already at variance with its powerful neighbour. The *Ch'un Ch'iu* first mentions Yüeh in 537, the *Tso Chuan* in 601.

[2] This is explicitly stated in the *Tso Chuan*, 昭公 XXXII, 2: 夏吳伐越始用師於越也.

B.C.

494	Fu Ch'ai defeats Kou Chien in the great battle of 夫 椒 Fu-chiao, and enters the capital of Yüeh.
485 or 484	Kou Chien renders homage to Wu. Death of Wu Tzŭ-hsü.
482	Kou Chien invades Wu in the absence of Fu Ch'ai.
478 476	Further attacks by Yüeh on Wu.
475	Kou Chien lays siege to the capital of Wu.
473	Final defeat and extinction of Wu.

The sentence quoted above from VI. § 21 hardly strikes me as one that could have been written in the full flush of victory. It seems rather to imply that, for the moment at least, the tide had turned against Wu, and that she was getting the worst of the struggle. Hence we may conclude that our treatise was not in existence in 505, before which date Yüeh does not appear to have scored any notable success against Wu. Ho Lu died in 496, so that if the book was written for him, it must have been during the period 505–496, when there was a lull in the hostilities, Wu having presumably been exhausted by its supreme effort against Ch'u. On the other hand, if we choose to disregard the tradition connecting Sun Wu's name with Ho Lu, it might equally well have seen the light between 496 and 494, or possibly in the period 482–473, when Yüeh was once again becoming a very serious menace. [1] We may feel fairly certain that the author, whoever he may have been, was not a man of any great eminence in his own day. On this point the negative testimony of the *Tso Chuan* far outweighs any shred of authority still attaching to the *Shih Chi*, if once its other facts are discredited. Sun Hsing-yen, however, makes a feeble attempt to explain the omission of his name from

---

[1] There is this to be said for the later period, that the feud would tend to grow more bitter after each encounter, and thus more fully justify the language used in XI. § 30.

the great commentary. It was Wu Tzŭ-hsü, he says, who got all the credit of Sun Wu's exploits, because the latter (being an alien) was not rewarded with an office in the State. [1]

How then did the Sun Tzŭ legend originate? It may be that the growing celebrity of the book imparted by degrees a kind of factitious renown to its author. It was felt to be only right and proper that one so well versed in the science of war should have solid achievements to his credit as well. Now the capture of Ying was undoubtedly the greatest feat of arms in Ho Lu's reign; it made a deep and lasting impression on all the surrounding states, and raised Wu to the short-lived zenith of her power. Hence, what more natural, as time went on, than that the acknowledged master of strategy, Sun Wu, should be popularly identified with that campaign, at first perhaps only in the sense that his brain conceived and planned it; afterwards, that it was actually carried out by him in conjunction with Wu Yüan, [2] Po P'ei and Fu Kai?

It is obvious that any attempt to reconstruct even the outline of Sun Tzŭ's life must be based almost wholly on conjecture. With this necessary proviso, I should say that he probably entered the service of Wu about the time of Ho Lu's accession, and gathered experience, though only in the capacity of a subordinate officer, during the intense military activity which marked the first half of that prince's reign. [3] If he rose to be a general at all, he certainly was never on an equal footing with the three

---

[1] See his preface to Sun Tzŭ: — 入郢威齊晉之功歸之子胥故春秋傳不載其名蓋功成不受官·

[2] With Wu Yüan himself the case is just the reverse: — a spurious treatise on war has been fathered on him simply because he was a great general. Here we have an obvious inducement to forgery. Sun Wu, on the other hand, cannot have been widely known to fame in the 5th century.

[3] See *Tso Chuan*, 定公, 4th year (506), § 14: 自昭王卽位無歲不有吳師 "From the date of King Chao's accession [515] there was no year in which Ch'u was not attacked by Wu."

above mentioned. He was doubtless present at the investment and occupation of Ying, and witnessed Wu's sudden collapse in the following year. Yüeh's attack at this critical juncture, when her rival was embarrassed on every side, seems to have convinced him that this upstart kingdom was the great enemy against whom every effort would henceforth have to be directed. Sun Wu was thus a well-seasoned warrior when he sat down to write his famous book, which according to my reckoning must have appeared towards the end, rather than the beginning, of Ho Lu's reign. The story of the women may possibly have grown out of some real incident occurring about the same time. As we hear no more of Sun Wu after this from any source, he is hardly likely to have survived his patron or to have taken part in the death-struggle with Yüeh, which began with the disaster at Tsui-li.

If these inferences are approximately correct, there is a certain irony in the fate which decreed that China's most illustrious man of peace should be contemporary with her greatest writer on war.

THE TEXT OF SUN TZŬ.

I have found it difficult to glean much about the history of Sun Tzŭ's text. The quotations that occur in early authors go to show that the "13 chapters" of which Ssŭ-ma Ch'ien speaks were essentially the same as those now extant. We have his word for it that they were widely circulated in his day, and can only regret that he refrained from discussing them on that account. [1] Sun Hsing-yen says in his preface : —

During the Ch'in and Han dynasties Sun Tzŭ's *Art of War* was in general use amongst military commanders, but they seem to have treated it as a work of mysterious import, and were unwilling to expound it for

---

[1] See *supra*, p. xx.

the benefit of posterity. Thus it came about that Wei Wu was the first to write a commentary on it. [1]

As we have already seen, there is no reasonable ground to suppose that Ts'ao Kung tampered with the text. But the text itself is often so obscure, and the number of editions which appeared from that time onward so great, especially during the T'ang and Sung dynasties, that it would be surprising if numerous corruptions had not managed to creep in. Towards the middle of the Sung period, by which time all the chief commentaries on Sun Tzŭ were in existence, a certain 吉天保 Chi T'ien-pao published a work in 15 *chüan* entitled 十家孫子會注 "Sun Tzŭ with the collected commentaries of ten writers." [2] There was another text, with variant readings put forward by Chu Fu of 大興 Ta-hsing, [3] which also had supporters among the scholars of that period; but in the Ming editions, Sun Hsing-yen tells us, these readings were for some reason or other no longer put into circulation. [4] Thus, until the end of the 18th century, the text in sole possession of the field was one derived from Chi T'ien-pao's edition, although no actual copy of that important work was known to have survived. That, therefore, is the text of Sun Tzŭ which appears in the War section of the great Imperial encyclopaedia printed in 1726, the 古今圖書集成 *Ku Chin T'u Shu Chi Ch'êng*. Another copy at my disposal of what is practically the same text, with slight variations, is that contained in the 周秦十一子 "Eleven philosophers of the Chou and Ch'in dynasties"

---

[1] 秦漢已來用兵皆用其法而或祕其書不肯注以傳世魏武始爲之注.

[2] See 宋藝文志.

[3] Alluded to on p. xvii, note 3.

[4] *Loc. cit.:* 蓋宋人又從大興朱氏處見明人刻本餘則世無傳者.

[1758]. And the Chinese printed in Capt. Calthrop's first edition is evidently a similar version which has filtered through Japanese channels. So things remained until 孫星衍 Sun Hsing-yen [1752–1818], a distinguished antiquarian and classical scholar,[1] who claimed to be an actual descendant of Sun Wu,[2] accidentally discovered a copy of Chi T'ien-pao's long-lost work, when on a visit to the library of the 華陰 Hua-yin temple.[3] Appended to it was the 遺說 *I Shuo* of 鄭友賢 Chêng Yu-hsien, mentioned in the *T'ung Chih*, and also believed to have perished.[4] This is what Sun Hsing-yen designates as the 古本 or 原本 "original edition (or text)" — a rather misleading name, for it cannot by any means claim to set before us the text of Sun Tzǔ in its pristine purity. Chi T'ien-pao was a careless compiler,[5] and appears to have been content to reproduce the somewhat debased version current in his day, without troubling to collate it

---

[1] A good biographical notice, with a list of his works, will be found in the 國朝詩人徵略, ch. 48, fol. 18 *sqq.*

[2] Preface *ad fin.:* 吾家出樂安眞孫子之後媿余徒讀祖書考証文字不通方略亦享承平之福者久也 "My family comes from Lo-an, and we are really descended from Sun Tzǔ. I am ashamed to say that I only read my ancestor's work from a literary point of view, without comprehending the military technique. So long have we been enjoying the blessings of peace!"

[3] Hua-yin is about 14 miles from 潼關 T'ung-kuan on the eastern border of Shensi. The temple in question is still visited by those about to make the ascent of the 華山 or Western Sacred Mountain. It is mentioned in the 大明一統志 [A.D. 1461], ch. 32, f. 22, as the 西嶽廟:— 在華陰縣東五里廟有唐玄宗所製華山碑 "Situated five *li* east of the district city of Hua-yin. The temple contains the Hua-shan tablet inscribed by the T'ang Emperor Hsüan Tsung [713—755]."

[4] 曩子游關中讀華陰嶽廟道藏見有此書後有鄭友賢遺說一卷.

[5] Cf. Sun Hsing-yen's remark *à propos* of his mistakes in the names and order of the commentators: 吉天保之不深究此書可知.

with the earliest editions then available. Fortunately, two versions of Sun Tzŭ, even older than the newly discovered work, were still extant, one buried in the *T'ung Tien*, Tu Yu's great treatise on the Constitution, the other similarly enshrined in the *T'ai P'ing Yü Lan* encyclopaedia. In both the complete text is to be found, though split up into fragments, intermixed with other matter, and scattered piecemeal over a number of different sections. Considering that the *Yü Lan* takes us back to the year 983, and the *T'ung Tien* about 200 years further still, to the middle of the T'ang dynasty, the value of these early transcripts of Sun Tzŭ can hardly be overestimated. Yet the idea of utilising them does not seem to have occurred to anyone until Sun Hsing-yen, acting under Government instructions, undertook a thorough recension of the text. This is his own account: —

> Because of the numerous mistakes in the text of Sun Tzŭ which his editors had handed down, the Government ordered that the ancient edition [of Chi T'ien-pao] should be used, and that the text should be revised and corrected throughout. It happened that Wu Nien-hu, the Governor Pi Kua, and Hsi, a graduate of the second degree, had all devoted themselves to this study, probably surpassing me therein. Accordingly, I have had the whole work cut on blocks as a text-book for military men. [1]

The three individuals here referred to had evidently been occupied on the text of Sun Tzŭ prior to Sun Hsing-yen's commission, but we are left in doubt as to the work they really accomplished. At any rate, the new edition, when ultimately produced, appeared in the names of Sun Hsing-yen and only one co-editor, 吳人驥 Wu Jên-chi. They took the "original text" as their basis, and by careful comparison with the older versions, as well as the extant commentaries and other sources of information such as

---

[1] 國家令甲以孫子校士所傳本或多錯謬當
用古本是正其文適吳念湖太守畢恬溪孝廉
皆爲此學所得或過于子遂刊一編以課武士.

the *I Shuo*, succeeded in restoring a very large number of doubtful passages, and turned out, on the whole, what must be accepted as the closest approximation we are ever likely to get to Sun Tzŭ's original work. This is what will hereafter be denominated the "standard text."

The copy which I have used belongs to a re-issue dated 1877. It is in 6 *pên*, forming part of a well-printed set of 23 early philosophical works in 83 *pên*.[1] It opens with a preface by Sun Hsing-yen (largely quoted in this introduction), vindicating the traditional view of Sun Tzŭ's life and performances, and summing up in remarkably concise fashion the evidence in its favour. This is followed by Ts'ao Kung's preface to his edition, and the biography of Sun Tzŭ from the *Shih Chi*, both translated above. Then come, firstly, Chêng Yu-hsien's *I Shuo*,[2] with author's preface, and next, a short miscellany of historical and bibliographical information entitled 孫子敍錄 *Sun Tzŭ Hsü Lu*, compiled by 畢以珣 Pi I-hsün. As regards the body of the work, each separate sentence is followed by a note on the text, if required, and then by the various commentaries appertaining to it, arranged in chronological order. These we shall now proceed to discuss briefly, one by one.

The Commentators.

Sun Tzŭ can boast an exceptionally long and distinguished roll of commentators, which would do honour to any classic. 歐陽修 Ou-yang Hsiu remarks on this fact, though he wrote before the tale was complete, and rather ingeniously explains it by saying that the artifices of war, being in-

---

[1] *See* my "Catalogue of Chinese Books" (Luzac & Co., 1908), no. 40.

[2] This is a discussion of 29 difficult passages in Sun Tzŭ, namely: I. 2; 20; 16; II. 9 & 10; III. 3; III & VII; III. 17; IV. 4; 6; V. 3; 10 & 11; 14; the headings of the 13 chapters, with special reference to chap. VII; VII. 5; 15 & 16; 27; 33, &c.; VIII. 1–6; IX. 11; X. 1–20; XI. 23; 31; 19; 43; VII. 12–14 & XI. 52; XI. 56; XIII. 15 & 16; 26; XIII in general.

exhaustible, must therefore be susceptible of treatment in a great variety of ways. [1]

1. 曹操 Ts'ao Ts'ao or 曹公 Ts'ao Kung, afterwards known as 魏武帝 Wei Wu Ti [A.D. 155–220]. There is hardly any room for doubt that the earliest commentary on Sun Tzŭ actually came from the pen of this extraordinary man, whose biography in the *San Kuo Chih* [2] reads like a romance. One of the greatest military geniuses that the world has seen, and Napoleonic in the scale of his operations, he was especially famed for the marvellous rapidity of his marches, which has found expression in the line 說曹操曹操就到 "Talk of Ts'ao Ts'ao, and Ts'ao Ts'ao will appear." Ou-yang Hsiu says of him that he was a great captain who "measured his strength against Tung Cho, Lü Pu and the two Yüan, father and son, and vanquished them all; whereupon he divided the Empire of Han with Wu and Shu, and made himself king. It is recorded that whenever a council of war was held by Wei on the eve of a far-reaching campaign, he had all his calculations ready; those generals who made use of them did not lose one battle in ten; those who ran counter to them in any particular saw their armies incontinently beaten and put to flight." [3] Ts'ao Kung's notes on Sun Tzŭ, models of austere brevity, are so thoroughly characteristic of the stern commander known to history, that it is hard indeed to conceive of them as the work of a mere *littérateur*. Sometimes, indeed, owing to extreme com-

---

[1] Preface to Mei Yao-ch'ên's edition: 孫子注者尤多武之書本於兵兵之術非一而以不窮爲奇宜其說者之多也. [2] See 魏書, ch. 1.

[3] *Loc. cit.*: 然前世言善用兵稱曹公曹公嘗與董呂諸袁角其力而勝之遂與吳蜀分漢而王傳言魏之將出兵千里每坐計勝敗授其成算諸將用之十不失一一有違者兵輒敗北.

pression, they are scarcely intelligible and stand no less in need of a commentary than the text itself.[1] As we have seen, Ts'ao Kung is the reputed author of the 新書, a book on war in 100,000 odd words, now lost, but mentioned in the 魏志.[2]

2. 孟氏 **Mêng Shih.** The commentary which has come down to us under this name is comparatively meagre, and nothing about the author is known. Even his personal name has not been recorded. Chi T'ien-pao's edition places him after Chia Lin, and 晁公武 Ch'ao Kung-wu also assigns him to the T'ang dynasty,[3] but this is obviously a mistake, as his work is mentioned in the 隋書經籍志. In Sun Hsing-yen's preface, he appears as Mêng Shih of the Liang dynasty [502–557]. Others would identify him with 孟康 Mêng K'ang of the 3rd century. In the 宋史藝文志,[4] he is named last of the 五家 "Five Commentators," the others being Wei Wu Ti, Tu Mu, Ch'ên Hao and Chia Lin.

3. 李筌 **Li Ch'üan** of the 8th century was a well-known writer on military tactics. His 太白陰經 has been in constant use down to the present day. The 通志 mentions 閫外春秋 (lives of famous generals from the Chou to the T'ang dynasty) as written by him.[5] He is also generally supposed to be the real author of the popular Taoist tract, the 陰符經. According to Ch'ao Kung-wu and the *T'ien-i-ko* catalogue,[6] he followed the 太乙遁甲 text of Sun Tzŭ, which differs considerably from those

---

[1] Cf. 天一閣藏書總目 Catalogue of the library of the 范 Fan family at Ningpo, 子部, fol. 12 *v*°: 其註多隱辭引而不發 "His commentary is frequently obscure; it furnishes a clue, but does not fully develop the meaning." [2] See 玉海, ch. 141 *ad init.*

[3] *Wên Hsien T'ung K'ao,* ch. 221, f. 9 *v*°. [4] Ch. 207, f. 5 *r*°.

[5] It is interesting to note that M. Pelliot has recently discovered chapters 1, 4 and 5 of this lost work in the "Grottos of the Thousand Buddhas." *See* B. E. F. E. O, t. VIII, nos. 3—4, p. 525. [6] *Loc. cit.*

now extant. His notes are mostly short and to the point, and he frequently illustrates his remarks by anecdotes from Chinese history.

4. 杜佑 **Tu Yu** (died 812) did not publish a separate commentary on Sun Tzŭ, his notes being taken from the *T'ung Tien*, the encyclopaedic treatise on the Constitution which was his life-work. They are largely repetitions of Ts'ao Kung and Mêng Shih, besides which it is believed that he drew on the ancient commentaries of 王凌 Wang Ling and others. Owing to the peculiar arrangement of the *T'ung Tien*, he has to explain each passage on its merits, apart from the context, and sometimes his own explanation does not agree with that of Ts'ao Kung, whom he always quotes first. Though not strictly to be reckoned as one of the "Ten Commentators," he was added to their number by Chi T'ien-pao, being wrongly placed after his grandson Tu Mu.

5. 杜牧 **Tu Mu** (803–852) is perhaps best known as a poet — a bright star even in the glorious galaxy of the T'ang period. We learn from Ch'ao Kung-wu that although he had no practical experience of war, he was extremely fond of discussing the subject, and was moreover well read in the military history of the *Ch'un Ch'iu* and *Chan Kuo* eras. [1] His notes, therefore, are well worth attention. They are very copious, and replete with historical parallels. The gist of Sun Tzŭ's work is thus summarised by him: "Practise benevolence and justice, but on the other hand make full use of artifice and measures of expediency." [2] He further declared that all the military

---

[1] *Wên Hsien T'ung K'ao*, ch. 221, f. 9: 世謂牧慨然最喜論兵欲試而不得者其學能道春秋戰國時事甚博而詳知兵者有取焉.

[2] Preface to his commentary (*T'u Shu*, 經籍典, ch. 442): 武之所論大約用仁義使機權也.

triumphs and disasters of the thousand years which had elapsed since Sun Wu's death would, upon examination, be found to uphold and corroborate, in every particular, the maxims contained in his book. [1] Tu Mu's somewhat spiteful charge against Ts'ao Kung has already been considered elsewhere.

6. 陳皥 **Ch'ên Hao** appears to have been a contemporary of Tu Mu. Ch'ao Kung-wu says that he was impelled to write a new commentary on Sun Tzŭ because Ts'ao Kung's on the one hand was too obscure and subtle, and that of Tu Mu on the other too long-winded and diffuse. [2] Ou-yang Hsiu, writing in the middle of the 11th century, calls Ts'ao Kung, Tu Mu and Ch'ên Hao the three chief commentators on Sun Tzŭ (三家), and observes that Ch'ên Hao is continually attacking Tu Mu's shortcomings. His commentary, though not lacking in merit, must rank below those of his predecessors.

7. 賈林 **Chia Lin** is known to have lived under the T'ang dynasty, for his commentary on Sun Tzŭ is mentioned in the 唐書 and was afterwards republished by 紀燮 Chi Hsieh of the same dynasty together with those of Mêng Shih and Tu Yu. [3] It is of somewhat scanty texture, and in point of quality, too, perhaps the least valuable of the eleven.

8. 梅堯臣 **Mei Yao-ch'ên** (1002–1060), commonly known by his "style" as Mei 聖俞 Shêng-yü, was, like Tu Mu, a poet of distinction. His commentary was published with a laudatory preface by the great Ou-yang Hsiu, from which we may cull the following: —

Later scholars have misread Sun Tzŭ, distorting his words and trying to make them square with their own one-sided views. Thus, though

---

[1] *Ibid.*: 自武死後凡千歲將兵者有成者有敗者勘其事跡皆與武所著書一一相抵當.

[2] *T'ung K'ao, loc. cit.*: 皥以曹公注隱微杜牧注闊疎重爲之注云.   [3] *Ibid.*

commentators have not been lacking, only a few have proved equal to the task. My friend Shêng-yü has not fallen into this mistake. In attempting to provide a critical commentary for Sun Tzŭ's work, he does not lose sight of the fact that these sayings were intended for states engaged in internecine warfare; that the author is not concerned with the military conditions prevailing under the sovereigns of the three ancient dynasties, [1] nor with the nine punitive measures prescribed to the Minister of War. [2] Again, Sun Wu loved brevity of diction, but his meaning is always deep. Whether the subject be marching an army, or handling soldiers, or estimating the enemy, or controlling the forces of victory, it is always systematically treated; the sayings are bound together in strict logical sequence, though this has been obscured by commentators who have probably failed to grasp their meaning. In his own commentary, Mei Shêng-yü has brushed aside all the obstinate prejudices of these critics, and has tried to bring out the true meaning of Sun Tzŭ himself. In this way, the clouds of confusion have been dispersed and the sayings made clear. I am convinced that the present work deserves to be handed down side by side with the three great commentaries; and for a great deal that they find in the sayings, coming generations will have constant reason to thank my friend Shêng-yü. [3]

Making some allowance for the exuberance of friendship, I am inclined to endorse this favourable judgment, and would certainly place him above Ch'ên Hao in order of merit.

---

[1] The Hsia, the Shang and the Chou. Although the last-named was nominally existent in Sun Tzŭ's day, it retained hardly a vestige of power, and the old military organisation had practically gone by the board. I can suggest no other explanation of the passage.

[2] See *Chou Li*, XXIX. 6–10.

[3] See *T'u Shu*, 戎政典, ch. 90, f. 2 *v°*:

後之學者徒見其
書又各牽於己見是以注者雖多而少當也獨之
吾友聖俞不然嘗評武之書曰此戰國相傾之及也
說也三代王者之師司馬九伐之法武不料敵制勝
然亦愛其文略而意深其行師用兵泪之或失其
亦皆有法其言甚有序次而注者泪之或失其意
意乃自爲注凡膠于偏見者皆抉去傳以己意與
而發之然後武之說不泪而明吾知此書常與
三家並傳而後世取其說者往往于吾聖俞多
焉.

9. 王晢 **Wang Hsi,** also of the Sung dynasty, is decidedly original in some of his interpretations, but much less judicious than Mei Yao-ch'ên, and on the whole not a very trustworthy guide. He is fond of comparing his own commentary with that of Ts'ao Kung, but the comparison is not often flattering to him. We learn from Ch'ao Kung-wu that Wang Hsi revised the ancient text of Sun Tzŭ, filling up lacunae and correcting mistakes. [1]

10. 何延錫 **Ho Yen-hsi** of the Sung dynasty. The personal name of this commentator is given as above by 鄭樵 Chêng Ch'iao in the *T'ung Chih*, written about the middle of the twelfth century, but he appears simply as 何氏 Ho Shih in the *Yü Hai*, and Ma Tuan-lin quotes Ch'ao Kung-wu as saying that his personal name is unknown. There seems to be no reason to doubt Chêng Ch'iao's statement, otherwise I should have been inclined to hazard a guess and identify him with one 何去非 Ho Ch'ü-fei, the author of a short treatise on war entitled 備論, who lived in the latter part of the 11th century. [2] Ho Shih's commentary, in the words of the *T'ien-i-ko* catalogue, 有所裨益 "contains helpful additions" here and there, but is chiefly remarkable for the copious extracts taken, in adapted form, from the dynastic histories and other sources.

11. 張預 **Chang Yü.** The list closes with a commentator of no great originality perhaps, but gifted with admirable powers of lucid exposition. His commentary is based on that of Ts'ao Kung, whose terse sentences he contrives to expand and develop in masterly fashion. Without Chang Yü, it is safe to say that much of Ts'ao Kung's commentary would have remained cloaked in its pristine obscurity and therefore valueless. His work is not mentioned in the Sung history, the *T'ung K'ao*, or

---

[1] *T'ung K'ao*, ch. 221, f. 11 r°: 晢以古本校正闕誤.
[2] See 四庫全書, ch. 99, f. 16 v°.

the *Yü Hai*, but it finds a niche in the *T'ung Chih*, which also names him as the author of the 百將傳 "Lives of Famous Generals." [1]

It is rather remarkable that the last-named four should all have flourished within so short a space of time. Ch'ao Kung-wu accounts for it by saying: "During the early years of the Sung dynasty the Empire enjoyed a long spell of peace, and men ceased to practise the art of war. But when [Chao] Yüan-hao's rebellion came [1038–42] and the frontier generals were defeated time after time, the Court made strenuous enquiry for men skilled in war, and military topics became the vogue amongst all the high officials. Hence it is that the commentators of Sun Tzŭ in our dynasty belong mainly to that period." [2]

Besides these eleven commentators, there are several others whose work has not come down to us. The *Sui Shu* mentions four, namely 王凌 Wang Ling (often quoted by Tu Yu as 王子); 張子尚 Chang Tzŭ-shang; 賈詡 Chia Hsü of 魏 Wei; [3] and 沈友 Shên Yu of 吳 Wu. The *T'ang Shu* adds 孫鎬 Sun Hao, and the *T'ung Chih* 蕭吉 Hsiao Chi, while the *T'u Shu* mentions a Ming commentator, 黃潤玉 Huang Jun-yü. It is possible that some of these may have been merely collectors and editors of other commentaries, like Chi T'ien-pao and Chi Hsieh, mentioned above. Certainly in the case of the latter, the entry 紀燮注孫子 in the *T'ung K'ao*, without the following note, would give one to understand that he had written an independent commentary of his own.

There are two works, described in the *Ssu K'u Ch'üan*

---

[1] This appears to be still extant. See Wylie's "Notes," p. 91 (new edition).

[2] *T'ung K'ao, loc. cit.*: 仁廟時天下久承平人不習兵元昊既叛邊將數敗朝廷頗訪知兵者士大夫人人言兵矣故本朝注解孫武書者大抵皆其時人也.

[3] A notable person in his day. His biography is given in the *San Kuo Chih*, ch. 10.

Shu [1] and no doubt extremely rare, which I should much like to have seen. One is entitled 孫子參同, in 5 *chüan*. It gives selections from four new commentators, probably of the Ming dynasty, as well as from the eleven known to us. The names of the four are 觧元 Hsieh Yüan; 張鰲 Chang Ao; 李村 Li Ts'ai; and 黃治徵 Huang Chih-chêng. The other work is 孫子彙徵 in 4 *chüan*, compiled by 鄭端 Chêng Tuan of the present dynasty. It is a compendium of information on ancient warfare, with special reference to Sun Tzŭ's 13 chapters.

APPRECIATIONS OF SUN TZŬ.

Sun Tzŭ has exercised a potent fascination over the minds of some of China's greatest men. Among the famous generals who are known to have studied his pages with enthusiasm may be mentioned 韓信 Han Hsin (*d.* B.C. 196), [2] 馮異 Fêng I (*d.* A.D. 34), [3] 呂蒙 Lü Mêng (*d.* 219), [4] and 岳飛 Yo Fei (1103–1141). [5] The opinion of Ts'ao Kung, who disputes with Han Hsin the highest place in Chinese military annals, has already been recorded. [6] Still more remarkable, in one way, is the testimony of purely literary men, such as 蘇洵 Su Hsün (the father of Su Tung-p'o), who wrote several essays on military topics, all of which owe their chief inspiration to Sun Tzŭ. The following short passage by him is preserved in the *Yü Hai:* [7] —

---

[1] Ch. 100, ff. 2, 3.   [2] *See* p. 144.   [3] *Hou Han Shu*, ch. 17 *ad init.*
[4] *San Kuo Chih*, ch. 54, f. 10 *v°* (commentary).
[5] *Sung Shih*, ch. 365 *ad init.*
[6] The few Europeans who have yet had an opportunity of acquainting themselves with Sun Tzŭ are not behindhand in their praise. In this connection, I may perhaps be excused for quoting from a letter from Lord Roberts, to whom the sheets of the present work were submitted previous to publication: "Many of Sun Wu's maxims are perfectly applicable to the present day, and no. 11 on page 77 is one that the people of this country would do well to take to heart."
[7] Ch. 140, f. 13 *r°*.

Sun Wu's saying, that in war one cannot make certain of conquering, [1] is very different indeed from what other books tell us. [2] Wu Ch'i was a man of the same stamp as Sun Wu: they both wrote books on war, and they are linked together in popular speech as "Sun and Wu." But Wu Ch'i's remarks on war are less weighty, his rules are rougher and more crudely stated, and there is not the same unity of plan as in Sun Tzŭ's work, where the style is terse, but the meaning fully brought out. [3]

The 性理彙娶, ch. 17, contains the following extract from the 藝圃折衷 "Impartial Judgments in the Garden of Literature" by 鄭厚 Chêng Hou: —

Sun Tzŭ's 13 chapters are not only the staple and base of all military men's training, but also compel the most careful attention of scholars and men of letters. His sayings are terse yet elegant, simple yet profound, perspicuous and eminently practical. Such works as the *Lun Yü*, the *I Ching* and the great Commentary, [4] as well as the writings of Mencius, Hsün K'uang and Yang Chu, all fall below the level of Sun Tzŭ. [5]

Chu Hsi, commenting on this, fully admits the first part of the criticism, although he dislikes the audacious comparison with the venerated classical works. Language of this sort, he says, "encourages a ruler's bent towards unrelenting warfare and reckless militarism." [6]

APOLOGIES FOR WAR.

Accustomed as we are to think of China as the greatest peace-loving nation on earth, we are in some danger of

---

[1] *See* IV. § 3.

[2] The allusion may be to Mencius VI. 2. ix. 2: 戰必克.

[3] 武用兵不能必克與書所言遠甚吳起與武一體之人皆著書言兵世稱之曰孫吳然而起之言兵也輕法制草略無所統紀不若武之書詞約而義盡.

[4] The *Tso Chuan*.

[5] 孫子十三篇不惟武人之根本文士亦當盡心焉其詞約而緐易而深暢而可用論語易大傳之流孟荀楊著書皆不及也.

[6] 是啟人君窮兵黷武之心.

forgetting that her experience of war in all its phases has also been such as no modern State can parallel. Her long military annals stretch back to a point at which they are lost in the mists of time. She had built the Great Wall and was maintaining a huge standing army along her frontier centuries before the first Roman legionary was seen on the Danube. What with the perpetual collisions of the ancient feudal States, the grim conflicts with Huns, Turks and other invaders after the centralisation of government, the terrific upheavals which accompanied the overthrow of so many dynasties, besides the countless rebellions and minor disturbances that have flamed up and flickered out again one by one, it is hardly too much to say that the clash of arms has never ceased to resound in one portion or another of the Empire.

No less remarkable is the succession of illustrious captains to whom China can point with pride. As in all countries, the greatest are found emerging at the most fateful crises of her history. Thus, Po Ch'i stands out conspicuous in the period when Ch'in was entering upon her final struggle with the remaining independent states. The stormy years which followed the break-up of the Ch'in dynasty are illumined by the transcendent genius of Han Hsin. When the House of Han in turn is tottering to its fall, the great and baleful figure of Ts'ao Ts'ao dominates the scene. And in the establishment of the T'ang dynasty, one of the mightiest tasks achieved by man, the superhuman energy of Li Shih-min (afterwards the Emperor T'ai Tsung) was seconded by the brilliant strategy of Li Ching. None of these generals need fear comparison with the greatest names in the military history of Europe.

In spite of all this, the great body of Chinese sentiment, from Lao Tzŭ downwards, and especially as reflected in the standard literature of Confucianism, has been consistently pacific and intensely opposed to militarism in any form. It is such an uncommon thing to find any of the literati defending warfare on principle, that I have thought it worth while to collect and translate a few passages in which the unorthodox view is upheld. The following, by Ssŭ-ma Ch'ien, shows that for all his ardent admiration of Confucius, he was yet no advocate of peace at any price: —

Military weapons are the means used by the Sage to punish violence and cruelty, to give peace to troublous times, to remove difficulties and dangers, and to succour those who are in peril. Every animal with blood in its veins and horns on its head will fight when it is attacked. How much more so will man, who carries in his breast the faculties of love and hatred, joy and anger! When he is pleased, a feeling of affection springs up within him; when angry, his poisoned sting is brought into play. That is the natural law which governs his being . . . . What then shall be said of those scholars of our time, blind to all great issues, and without any appreciation of relative values, who can only bark out their stale formulas about "virtue" and "civilisation," condemning the use of military weapons? They will surely bring our country to impotence and dishonour and the loss of her rightful heritage; or, at the very least, they will bring about invasion and rebellion, sacrifice of territory and general enfeeblement. Yet they obstinately refuse to modify the position they have taken up. The truth is that, just as in the family the teacher must not spare the rod, and punishments cannot be dispensed with in the State, so military chastisement can never be allowed to fall into abeyance in the Empire. All one can say is that this power will be exercised wisely by some, foolishly by others, and that among those who bear arms some will be loyal and others rebellious. [1]

The next piece is taken from Tu Mu's preface to his commentary on Sun Tzŭ: —

War may be defined as punishment, which is one of the functions of government. It was the profession of Chung Yu and Jan Ch'iu, both

---

[1] *Shih Chi*, ch. 25, fol. 1: 兵者聖人所以討彊暴平亂
世夷險阻救危殆自含血戴角之獸見犯則校
而況於人懷好惡喜怒之氣喜則愛心生怒則
毒螫加情性之理也⋯豈與世儒闇於大較不
權輕重猥云德化不當用兵大至奢辱失守小
乃侵犯削弱遂執不移等哉故教笞不可廢於
家刑罰不可捐於國誅伐不可偃於天下用之
有巧拙行之有逆順耳.

disciples of Confucius. Nowadays, the holding of trials and hearing of litigation, the imprisonment of offenders and their execution by flogging in the market-place, are all done by officials. But the wielding of huge armies, the throwing down of fortified cities, the haling of women and children into captivity, and the beheading of traitors — this is also work which is done by officials. The objects of the rack [1] and of military weapons are essentially the same. There is no intrinsic difference between the punishment of flogging and cutting off heads in war. For the lesser infractions of law, which are easily dealt with, only a small amount of force need be employed: hence the institution of torture and flogging. For more serious outbreaks of lawlessness, which are hard to suppress, a greater amount of force is necessary: hence the use of military weapons and wholesale decapitation. In both cases, however, the end in view is to get rid of wicked people, and to give comfort and relief to the good [2] ....

Chi-sun asked Jan Yu, saying: "Have you, Sir, acquired your military aptitude by study, or is it innate?" Jan Yu replied: "It has been acquired by study." [3] "How can that be so," said Chi-sun, "seeing that you are a disciple of Confucius?" "It is a fact," replied Jan Yu; "I was taught by Confucius. It is fitting that the great Sage should exercise both civil and military functions, though to be sure my instruction in the art of fighting has not yet gone very far."

Now, who the author was of this rigid distinction between the "civil" and the "military," and the limitation of each to a separate sphere of action, or in what year of which dynasty it was first introduced, is more than I can say. But, at any rate, it has come about that the members of the governing class are quite afraid of enlarging on military topics, or do so only in a shamefaced manner. If any are bold enough to discuss the subject, they are at once set down as eccentric individuals of coarse and brutal propensities. This is an extraordinary instance of the way in

---

[1] The first instance of 木索 given in the *P'ei Wên Yün Fu* is from Ssŭ-ma Ch'ien's letter to 任安 Jên An (see 文選, ch. 41, f. 9 r°), where M. Chavannes translates it "la cangue et la chaîne." But in the present passage it seems rather to indicate some single instrument of torture.

[2] 兵者刑也刑者政事也爲夫于之徒實仲由
冉求之事也今者據案聽訟械繫罪人笞死于
市者吏之所爲也驅兵數萬撅其城郭虜其妻
于斬其罪人亦吏之所爲也木索兵刃無異意
也笞之與斬無異刑也小而易制用力少者木
索笞也大而難治用力多者兵刃斬也俱期於
除去惡民安活善民·

[3] Cf. *Shih Chi*, ch. 47, f. 11 v°.

which, through sheer lack of reasoning, men unhappily lose sight of fundamental principles.

When the Duke of Chou was minister under Ch'êng Wang, he regulated ceremonies and made music, and venerated the arts of scholarship and learning; yet when the barbarians of the River Huai revolted, [2] he sallied forth and chastised them. When Confucius held office under the Duke of Lu, and a meeting was convened at Chia-ku, [3] he said: "If pacific negotiations are in progress, warlike preparations should have been made beforehand." He rebuked and shamed the Marquis of Ch'i, who cowered under him and dared not proceed to violence. How can it be said that these two great Sages had no knowledge of military matters? [4]

We have seen that the great Chu Hsi held Sun Tzǔ in high esteem. He also appeals to the authority of the Classics: —

Our Master Confucius, answering Duke Ling of Wei, said: "I have never studied matters connected with armies and battalions." [5] Replying to K'ung Wên-tzǔ, he said: "I have not been instructed about buff-coats and weapons." [6] But if we turn to the meeting at Chia-ku, [7] we find that he used armed force against the men of Lai, [8] so that the marquis of Ch'i was overawed. Again, when the inhabitants of Pi revolted, he ordered his officers to attack them, whereupon they were defeated and fled in confusion. [9] He once uttered the words: "If I fight, I con-

---

[1] 季孫問于冉有曰子之戰學之乎性達之乎
對曰學之季孫曰事孔子惡乎學冉有曰卽學
之於孔子者大聖兼該文武並用適聞其戰法
實未之詳也夫不知自何代何年何人分爲二
道曰文曰武離而俱行因使縉紳之士不敢言
兵甚或恥言之苟有言者世以爲麤暴異人人
不比數嗚呼亡失根本斯爲最甚·

[2] See *Shu Ching*, preface § 55.

[3] See *Tso Chuan*, 定公 X. 2; *Shih Chi*, ch. 47, f. 4 r°.

[4] 周公相成王制禮作樂尊大儒術有淮夷叛
則出征之夫子相魯公會于夾谷曰有文事者
必有武備比辱齊侯伏不敢動是二大聖人豈
不知兵乎·

[5] *Lun Yü*, XV. 1.

[6] *Tso Chuan*, 哀公, XI. 7.

[7] See *supra*.

[8] *Tso Chuan*, 定公, X. 2.

[9] *Ibid*. XII. 5; *Chia Yü*, ch. 1 *ad fin*.

quer." [1]   And Jan Yu also said: "The Sage exercises both civil and military functions." [2]   Can it be a fact that Confucius never studied or received instruction in the art of war? We can only say that he did not specially choose matters connected with armies and fighting to be the subject of his teaching. [3]

Sun Hsing-yen, the editor of Sun Tzŭ, writes in similar strain: —

Confucius said: "I am unversed in military matters." [4]   He also said: "If I fight, I conquer." [4]   Confucius ordered ceremonies and regulated music. Now war constitutes one of the five classes of State ceremonial, [5] and must not be treated as an independent branch of study. Hence, the words "I am unversed in" must be taken to mean that there are things which even an inspired Teacher does not know. Those who have to lead an army and devise stratagems, must learn the art of war. But if one can command the services of a good general like Sun Tzŭ, who was employed by Wu Tzŭ-hsü, there is no need to learn it oneself. Hence the remark added by Confucius: "If I fight, I conquer." [6]

The men of the present day, however, wilfully interpret these words of Confucius in their narrowest sense, as though he meant that books on the art of war were not worth reading. With blind persistency, they adduce the example of Chao Kua, who pored over his father's books to no purpose, [7] as a proof that all military theory is useless. Again, seeing

---

[1] have failed to trace this utterance. See note 2 on p. xliii.

[2] See *supra.*

[3] 性理彙要, *loc. cit.*: 昔吾夫子對衞靈公以軍旅之事未之學答孔文子以甲兵之事未之聞及觀夾谷之會則以兵加萊人而齊侯懼費人之亂則命將士以伐之而費人北嘗曰我戰則克而冉有亦曰聖人文武並用孔子豈有眞未學未聞哉特以軍旅甲兵之事非所以爲訓也.

[4] See *supra.*

[5] *Viz.,* 軍禮, the other four being 吉, 凶, 賓 and 嘉 "worship, mourning, entertainment of guests and festive rites." See *Shu Ching,* II. 1. iii. 8, and *Chou Li,* IX. fol. 49.

[6] Preface to Sun Tzŭ: 孔子曰軍旅之事未之學又曰我戰則克孔子定禮正樂兵則五禮之一不必以爲專門之學故云未學所爲聖人有所不知或行軍好謀則學之或善將將如伍子胥之用孫子又何必自學之故又曰我戰則克也,

[7] See p. 166.

that books on war have to do with such things as opportunism in design-
ing plans, and the conversion of spies, they hold that the art is immoral
and unworthy of a sage.　These people ignore the fact that the studies
of our scholars and the civil administration of our officials also require
steady application and practice before efficiency is reached.　The ancients
were particularly chary of allowing mere novices to botch their work. [1]
Weapons are baneful [2] and fighting perilous; and unless a general is in
constant practice, he ought not to hazard other men's lives in battle. [3]
Hence it is essential that Sun Tzŭ's 13 chapters should be studied. [4]

Hsiang Liang used to instruct his nephew Chi [5] in the art of war.
Chi got a rough idea of the art in its general bearings, but would not
pursue his studies to their proper outcome, the consequence being that
he was finally defeated and overthrown.　He did not realise that the
tricks and artifices of war are beyond verbal computation.　Duke Hsiang
of Sung [6] and King Yen of Hsü [7] were brought to destruction by their
misplaced humanity.　The treacherous and underhand nature of war
necessitates the use of guile and stratagem suited to the occasion.　There
is a case on record of Confucius himself having violated an extorted
oath, [8] and also of his having left the Sung State in disguise. [9]　Can we
then recklessly arraign Sun Tzŭ for disregarding truth and honesty? [10]

---

[1] This is a rather obscure allusion to *Tso Chuan*, 襄公, XXXI. 4, where

Tzŭ-ch'an says: 子有美錦不使人學製焉 "If you have a piece
of beautiful brocade, you will not employ a mere learner to make it up."

[2] Cf. *Tao Tê Ching*, ch. 31: 兵者不祥之器.

[3] Sun Hsing-yen might have quoted Confucius again. See *Lun Yü*, XIII. 29, 30.

[4] 今世泥孔子之言以爲兵書不足觀又泥趙
括徒能讀父書之言以爲成法不足用又見兵
書有權謀有反間以爲非聖人之法皆不知吾
儒之學者吏之治事可習而能然古人猶有學
製之懼兵凶戰危將不素習未可以人命爲嘗
試則十三篇之不可不觀也.

[5] Better known as Hsiang 羽 Yü [B.C. 233–202].

[6] The third among the 五伯 (or 霸) enumerated on p. 141.　For the incident referred to, see *Tso Chuan*, 僖公, XXII. 4.

[7] See *supra*, p. xvi, note 4.　　　　[8] *Shih Chi*, ch. 47, f. 7 *r*°.

[9] *Ibid.*, ch. 38, f. 8 *v*°.

[10] 項梁教籍兵法籍略知其意不肯竟學卒以
傾覆不知兵法之弊可勝言哉宋襄徐偃仁而

BIBLIOGRAPHY.

The following are the oldest Chinese treatises on war, after Sun Tzŭ. The notes on each have been drawn principally from the 四庫全書簡明目錄 *Ssŭ k'u ch'üan shu chien ming mu lu*, ch. 9, fol. 22 *sqq.*

1. 吳子 **Wu Tzŭ,** in 1 *chüan* or 6 篇 chapters. By 吳起 Wu Ch'i (*d.* B.C. 381). A genuine work. See *Shih Chi*, ch. 65.

2. 司馬法 **Ssŭ-ma Fa,** in 1 *chüan* or 5 chapters. Wrongly attributed to 司馬穰苴 Ssŭ-ma Jang-chü of the 6th century B.C. Its date, however, must be early, as the customs of the three ancient dynasties are constantly to be met with in its pages. [1] See *Shih Chi*, ch. 64.

The *Ssŭ K'u Ch'üan Shu* (ch. 99, f. 1) remarks that the oldest three treatises on war, *Sun Tzŭ*, *Wu Tzŭ* and the *Ssŭ-ma Fa*, are, generally speaking, only concerned with things strictly military — the art of producing, collecting, training and drilling troops, and the correct theory with regard to measures of expediency, laying plans, transport of goods and the handling of soldiers [2] — in strong contrast to later works, in which the science of war is usually blended with metaphysics, divination and magical arts in general.

3. 六韜 **Liu T'ao,** in 6 *chüan* or 60 chapters. Attributed to 呂望 Lü Wang (or Lü 尚 Shang, also known as 太公 T'ai Kung) of the 12th century B.C. [3] But

---

敗兵者危機當用權謀孔子猶有要盟勿信微
服過宋之時安得妄責孫子以言之不純哉·
[1] 其時去古未遠三代遺規往往於此書見之·
[2] 其最古者當以孫子吳子司馬法爲本大抵
生聚訓練之術權謀運用之宜而已·

[3] See p. 174. Further details on T'ai Kung will be found in the *Shih Chi*, ch. 32 *ad init.* Besides the tradition which makes him a former minister of Chou Hsin, two other accounts of him are there given, according to which he would appear to have been first raised from a humble private station by Wên Wang.

its style does not belong to the era of the Three Dynasties. [1] 陸德明 Lu Tê-ming (550–625 A.D.) mentions the work, and enumerates the headings of the six sections, 文, 武, 虎, 豹, 龍 and 犬, so that the forgery cannot have been later than the Sui dynasty.

4. 尉繚子 **Wei Liao Tzǔ**, in 5 *chüan*. Attributed to Wei Liao (4th cent. B.C.), who studied under the famous 鬼谷子 Kuei-ku Tzǔ. The 漢志, under 兵家, mentions a book of Wei Liao in 31 chapters, whereas the text we possess contains only 24. Its matter is sound enough in the main, though the strategical devices differ considerably from those of the Warring States period. [2] It has been furnished with a commentary by the well-known Sung philosopher 張載 Chang Tsai.

5. 三略 **San Lüeh**, in 3 *chüan*. Attributed to 黄石公 Huang-shih Kung, a legendary personage who is said to have bestowed it on Chang Liang (*d.* B.C. 187) in an interview on a bridge. [3] But here again, the style is not that of works dating from the Ch'in or Han period. The Han Emperor Kuang Wu [A.D. 25–57] apparently quotes from it in one of his proclamations; but the passage in question may have been inserted later on, in order to prove the genuineness of the work. We shall not be far out if we refer it to the Northern Sung period [420–478 A.D.], or somewhat earlier. [4]

---

[1] 其文義不類三代.

[2] 其言多近於正與戰國權謀頗殊.

[3] See *Han Shu*, 張良傳, ch. 40. The work is there called 太公兵法. Hence it has been confused with the *Liu T'ao*. The *T'u Shu* attributes both the *Liu T'ao* and the *San Lüeh* to T'ai Kung.

[4] 其文不類秦漢間書漢光武帝詔雖嘗引之安知非反憮詔中所引二語以證實其書謂之北宋以前舊本則可矣. Another work said to have been written by Huang-shih Kung, and also included in the military section of the Imperial Catalogue, is the 素書 *Su Shu* in 1 *chüan*. A short ethical treatise of Taoist

6. 李衞公問對 **Li Wei Kung Wên Tui,** in 3 sections. Written in the form of a dialogue between T'ai Tsung and his great general 李靖 Li Ching, it is usually ascribed to the latter. Competent authorities consider it a forgery, though the author was evidently well versed in the art of war. [1]

7. 李靖兵法 **Li Ching Ping Fa** (not to be confounded with the foregoing) is a short treatise in 8 chapters, preserved in the *T'ung Tien*, but not published separately. This fact explains its omission from the *Ssŭ K'u Ch'üan Shu.*

8. 握奇經 **Wu Ch'i Ching,** [2] in 1 *chüan*. Attributed to the legendary minister 風后 Fêng Hou, with exegetical notes by 公孫宏 Kung-sun Hung of the Han dynasty (*d.* B.C. 121), and said to have been eulogised by the celebrated general 馬隆 Ma Lung (*d.* A.D. 300). Yet the earliest mention of it is in the 宋志. Although a forgery, the work is well put together. [3]

Considering the high popular estimation in which 諸 葛亮 Chu-ko Liang has always been held, it is not surprising to find more than one work on war ascribed to his pen. Such are (1) the 十六策 **Shih Liu Ts'ê** (1 *chüan*), preserved in the 永樂大典 *Yung Lo Ta Tien;* (2) 將苑 **Chiang Yüan** (1 *ch.*); and (3) 心書 **Hsin Shu** (1 *ch.*), which steals wholesale from Sun Tzŭ. None of these has the slightest claim to be considered genuine.

---

savour, having no reference whatever to war, it is pronounced a forgery from the hand of 張商英 Chang Shang-ying (*d.* 1121), who edited it with commentary. Correct Wylie's "Notes," new edition, p. 90, and Courant's "Catalogue des Livres Chinois," no. 5056.

[1] 其書雖僞亦出於有學識謀略者之手也. We are told in the 讀書志 that the above six works, together with Sun Tzŭ, were those prescribed for military training in the 元豐 period (1078–85). See *Yü Hai*, ch. 140, f. 4 *r*°.

[2] Also written 握機經 and 幄機經 *Wu Chi Ching.*

[3] 其言具有條理.

Most of the large Chinese encyclopaedias contain extensive sections devoted to the literature of war. The following references may be found useful: —

通典 T'ung Tien (*circâ* 800 A.D.), ch. 148–162.

太平御覽 T'ai P'ing Yü Lan (983), ch. 270–359.

文獻通考 Wên Hsien T'ung K'ao (13th cent.), ch. 221.

玉海 Yü Hai (13th cent.), ch. 140, 141.

三才圖會 San Ts'ai T'u Hui (16th cent.), 人事 ch. 7, 8.

廣博物志 Kuang Po Wu Chih (1607), ch. 31, 32.

潛確類書 Ch'ien Ch'io Lei Shu (1632), ch. 75.

淵鑑類函 Yüan Chien Lei Han (1710), ch. 206–229.

古今圖書集成 Ku Chin T'u Shu Chi Ch'êng (1726), section XXX, *esp.* ch. 81–90.

續文獻通考 Hsü Wên Hsien T'ung K'ao (1784), ch. 121–134.

皇朝經世文編 Huang Ch'ao Ching Shih Wên Pien (1826), ch. 76, 77.

The bibliographical sections of certain historical works also deserve mention: —

前漢書 Ch'ien Han Shu, ch. 30.

隋書 Sui Shu, ch. 32–35.

舊唐書 Chiu T'ang Shu, ch. 46, 47.

新唐書 Hsin T'ang Shu, ch. 57–60.

宋史 Sung Shih, ch. 202–209.

通志 T'ung Chih (*circâ* 1150), ch. 68.

To these of course must be added the great Catalogue of the Imperial Library: —

四庫全書總目提要 Ssǔ K'u Ch'üan Shu Tsung Mu T'i Yao (1790), ch. 99, 100.

I. 計篇.

1. 孫子曰兵者國之大事
2. 死生之地存亡之道不可不察也
3. 故經之以五校之以計而索其情

I. LAYING PLANS.

This is the only possible meaning of 計, which M. Amiot and Capt. Calthrop wrongly translate "Fondements de l'art militaire" and "First principles" respectively. Ts'ao Kung says it refers to the deliberations in the temple selected by the general for his temporary use, or as we should say, in his tent. See § 26.

1. Sun Tzǔ said: The art of war is of vital importance to the State.

2. It is a matter of life and death, a road either to safety or to ruin. Hence it is a subject of inquiry which can on no account be neglected.

3. The art of war, then, is governed by five constant factors, to be taken into account in one's deliberations, when seeking to determine the conditions obtaining in the field.

The old text of the *T'ung Tien* has 故經之以五校之計, etc. Later editors have inserted 事 after 五, and 以 before 計. The former correction is perhaps superfluous, but the latter seems necessary in order to make sense, and is supported by the accepted reading in § 12, where the same words recur. I am inclined to think, however, that the whole sentence from 校 to 情 is an interpolation and has no business here at all. If it be retained, Wang Hsi must be right in saying that 計 denotes the "seven considerations" in § 13. 情 are the circumstances or conditions likely to bring about victory or defeat. The antecedent of the first 之 is 兵者; of the second, 五. 校

4. 一曰道二曰天三曰地四曰將五曰法
5. 道者令民與上同意也
6. 故可與之死可與之生而民不畏危
7. 天者陰陽寒暑時制也
8. 地者遠近險易廣狹死生也

contains the idea of "comparison with the enemy," which cannot well be brought out here, but will appear in § 12. Altogether, difficult though it is, the passage is not so hopelessly corrupt as to justify Capt. Calthrop in burking it entirely.

4. These are: (1) The Moral Law; (2) Heaven; (3) Earth; (4) The Commander; (5) Method and discipline.

It appears from what follows that Sun Tzŭ means by 道 a principle of harmony, not unlike the Tao of Lao Tzŭ in its moral aspect. One might be tempted to render it by "morale," were it not considered as an attribute of the *ruler* in § 13.

5, 6. *The Moral Law* causes the people to be in complete accord with their ruler, so that they will follow him regardless of their lives, undismayed by any danger.

The original text omits 令民, inserts an 以 after each 可, and omits 民 after 而. Capt. Calthrop translates: "If the ruling authority be upright, the people are united" — a very pretty sentiment, but wholly out of place in what purports to be a translation of Sun Tzŭ.

7. *Heaven* signifies night and day, cold and heat, times and seasons.

The commentators, I think, make an unnecessary mystery of 陰陽. Thus Mêng Shih defines the words as 剛柔盈縮 "the hard and the soft, waxing and waning," which does not help us much. Wang Hsi, however, may be right in saying that what is meant is 總天道 "the general economy of Heaven," including the five elements, the four seasons, wind and clouds, and other phenomena.

8. *Earth* comprises distances, great and small; danger and security; open ground and narrow passes; the chances of life and death.

死生 (omitted by Capt. Calthrop) may have been included here because the safety of an army depends largely on its quickness to turn these geographical features to account.

9. 將者智信仁勇嚴也
10. 法者曲制官道主用也
11. 凡此五者將莫不聞知之者勝不知者不勝
12. 故校之以計而索其情

9. *The Commander* stands for the virtues of wisdom, sincerity, benevolence, courage and strictness.

The five cardinal virtues of the Chinese are (1) 仁 humanity or benevolence; (2) 義 uprightness of mind; (3) 禮 self-respect, self-control, or "proper feeling;" (4) 智 wisdom; (5) 信 sincerity or good faith. Here 智 and 信 are put before 仁, and the two military virtues of "courage" and "strictness" substituted for 義 and 禮.

10. By *Method and discipline* are to be understood the marshalling of the army in its proper subdivisions, the gradations of rank among the officers, the maintenance of roads by which supplies may reach the army, and the control of military expenditure.

The Chinese of this sentence is so concise as to be practically unintelligible without commentary. I have followed the interpretation of Ts'ao Kung, who joins 曲 制 and again 主 用. Others take each of the six predicates separately. 曲 has the somewhat uncommon sense of "cohort" or division of an army. Capt. Calthrop translates: "Partition and ordering of troops," which only covers 曲 制.

11. These five heads should be familiar to every general: he who knows them will be victorious; he who knows them not will fail.

12. Therefore, in your deliberations, when seeking to determine the military conditions, let them be made the basis of a comparison, in this wise: —

The *Yü Lan* has an interpolated 五 before 計. It is obvious, however, that the 五 者 just enumerated cannot be described as 計. Capt. Calthrop, forced to give some rendering of the words which he had omitted in § 3, shows himself decidedly hazy: "Further, with regard to these and the following seven matters, the condition of the enemy must be compared with our own." He does not appear to see that the seven queries or considerations which follow arise directly out of the Five heads, instead of being supplementary to them.

13. 日主孰有道將孰有能天地孰得法令孰行
兵眾孰強士卒孰練賞罰孰明

14. 吾以此知勝負矣

13. (1) Which of the two sovereigns is imbued with the Moral law?

I. e., "is in harmony with his subjects." Cf. § 5.

(2) Which of the two generals has most ability?

(3) With whom lie the advantages derived from Heaven and Earth?

See §§ 7, 8.

(4) On which side is discipline most rigorously enforced?

Tu Mu alludes to the remarkable story of Ts'ao Ts'ao (A. D. 155—220), who was such a strict disciplinarian that once, in accordance with his own severe regulations against injury to standing crops, he condemned himself to death for having allowed his horse to shy into a field of corn! However, in lieu of losing his head, he was persuaded to satisfy his sense of justice by cutting off his hair. Ts'ao Ts'ao's own comment on the present passage is characteristically curt: 設而不犯犯而必誅 "when you lay down a law, see that it is not disobeyed; if it is disobeyed, the offender must be put to death."

(5) Which army is the stronger?

Morally as well as physically. As Mei Yao-ch'ên puts it, 內和外附, which might be freely rendered "esprit de corps and 'big battalions.'"

(6) On which side are officers and men more highly trained?

Tu Yu quotes 王子 as saying: "Without constant practice, the officers will be nervous and undecided when mustering for battle; without constant practice, the general will be wavering and irresolute when the crisis is at hand."

(7) In which army is there the greater constancy both in reward and punishment?

明, literally "clear;" that is, on which side is there the most absolute certainty that merit will be properly rewarded and misdeeds summarily punished?

14. By means of these seven considerations I can forecast victory or defeat.

15. 將聽吾計用之必勝留之將不聽吾計用之
必敗去之

16. 計利以聽乃爲之勢以佐其外

17. 勢者因利而制權也

15. The general that hearkens to my counsel and acts upon it, will conquer: — let such a one be retained in command! The general that hearkens not to my counsel nor acts upon it, will suffer defeat: — let such a one be dismissed!

The form of this paragraph reminds us that Sun Tzŭ's treatise was composed expressly for the benefit of his patron 闔 閭 Ho Lü, king of the Wu State. It is not necessary, however, to understand 我 before 留 之 (as some commentators do), or to take 將 as "generals under my command."

16. While heeding the profit of my counsel, avail yourself also of any helpful circumstances over and beyond the ordinary rules.

Capt. Calthrop blunders amazingly over this sentence: "Wherefore, with regard to the foregoing, considering that with us lies the advantage, and the generals agreeing, we create a situation which promises victory." Mere logic should have kept him from penning such frothy balderdash.

17. According as circumstances are favourable, one should modify one's plans.

Sun Tzŭ, as a practical soldier, will have none of the "bookish theoric." He cautions us here not to pin our faith to abstract principles; "for," as Chang Yü puts it, "while the main laws of strategy can be stated clearly enough for the benefit of all and sundry, you must be guided by the actions of the enemy in attempting to secure a favourable position in actual warfare." On the eve of the battle of Waterloo, Lord Uxbridge, commanding the cavalry, went to the Duke of Wellington in order to learn what his plans and calculations were for the morrow, because, as he explained, he might suddenly find himself Commander-in-chief and would be unable to frame new plans in a critical moment. The Duke listened quietly and then said: "Who will attack the first to-morrow — I or Bonaparte?" "Bonaparte," replied Lord Uxbridge. "Well," continued the Duke, "Bonaparte has not given me any idea of his projects; and as my plans will depend upon his, how can you expect me to tell you what mine are?" *

---

* "Words on Wellington," by Sir W. Fraser.

18. 兵者詭道也

19. 故能而示之不能用而示之不用近而示之
遠遠而示之近

20. 利而誘之亂而取之

21. 實而備之強而避之

22. 怒而撓之卑而驕之

18. All warfare is based on deception.

The truth of this pithy and profound saying will be admitted by every soldier. Col. Henderson tells us that Wellington, great in so many military qualities, was especially distinguished by "the extraordinary skill with which he concealed his movements and deceived both friend and foe."

19. Hence, when able to attack, we must seem unable; when using our forces, we must seem inactive; when we are near, we must make the enemy believe we are far away; when far away, we must make him believe we are near.

20. Hold out baits to entice the enemy. Feign disorder, and crush him.

取, as often in Sun Tzŭ, is used in the sense of 擊. It is rather remarkable that all the commentators, with the exception of Chang Yü, refer 亂 to the enemy: "when he is in disorder, crush him." It is more natural to suppose that Sun Tzŭ is still illustrating the uses of deception in war.

21. If he is secure at all points, be prepared for him. If he is in superior strength, evade him.

The meaning of 實 is made clear from chap. VI, where it is opposed to 虛 "weak or vulnerable spots." 強, according to Tu Yu and other commentators, has reference to the keenness of the men as well as to numerical superiority. Capt. Calthrop evolves an extraordinarily far-fetched translation: "If there are defects, give an appearance of perfection, and awe the enemy. Pretend to be strong, and so cause the enemy to avoid you"!

22. If your opponent is of choleric temper, seek to irritate him. Pretend to be weak, that he may grow arrogant.

I follow Chang Yü in my interpretation of 怒. 卑 is expanded by Mei Yao-ch'ên into 示以卑弱. Wang Tzŭ, quoted by Tu Yu,

23. 佚而勞之親而離之
24. 攻其無備出其不意
25. 此兵家之勝不可先傳也
26. 夫未戰而廟算勝者得算多也未戰而廟算
不勝者得算少也多算勝少算不勝而況於
無算乎吾以此觀之勝負見矣

says that the good tactician plays with his adversary as a cat plays with a mouse, first feigning weakness and immobility, and then suddenly pouncing upon him.

23. If he is taking his ease, give him no rest.

This is probably the meaning, though Mei Yao-ch'ên has the note: 以我之佚待彼之勞 "while we are taking our ease, wait for the enemy to tire himself out." The *Yü Lan* has 引而勞之 "Lure him on and tire him out." This would seem also to have been Ts'ao Kung's text, judging by his comment 以利勞之.

If his forces are united, separate them.

Less plausible is the interpretation favoured by most of the commentators: "If sovereign and subject are in accord, put division between them."

24. Attack him where he is unprepared, appear where you are not expected.

25. These military devices, leading to victory, must not be divulged beforehand.

This seems to be the way in which Ts'ao Kung understood the passage, and is perhaps the best sense to be got out of the text as it stands. Most of the commentators give the following explanation: "It is impossible to lay down rules for warfare before you come into touch with the enemy." This would be very plausible if it did not ignore 此, which unmistakably refers to the maxims which Sun Tzǔ *has* been laying down. It is possible, of course, that 此 may be a later interpolation, in which case the sentence would practically mean: "Success in warfare cannot be taught." As an alternative, however, I would venture to suggest that a second 不 may have fallen out after 可, so that we get: "These maxims for succeeding in war are the first that ought to be imparted."

26. Now the general who wins a battle makes many calculations in his temple ere the battle is fought.

Chang Yü tells us that in ancient times it was customary for a temple to be set apart for the use of a general who was about to take the field, in order that he might there elaborate his plan of campaign. Capt. Calthrop misunderstands it as "the shrine of the ancestors," and gives a loose and inaccurate rendering of the whole passage.

The general who loses a battle makes but few calculations beforehand. Thus do many calculations lead to victory, and few calculations to defeat: how much more no calculation at all! It is by attention to this point that I can foresee who is likely to win or lose.

II. 作戰篇.

1. 孫于曰凡用兵之法馳車千駟革車千乘帶
甲十萬千里饋糧則內外之費賓客之用膠
漆之材車甲之奉日費千金然後十萬之師
舉矣

II. WAGING WAR.

Ts'ao Kung has the note: 欲戰必先算其費務 "He who wishes to fight must first count the cost," which prepares us for the discovery that the subject of the chapter is not what we might expect from the title, but is primarily a consideration of ways and means.

1. Sun Tzŭ said: In the operations of war, where there are in the field a thousand swift chariots, as many heavy chariots, and a hundred thousand mail-clad soldiers,

The 馳車 were lightly built and, according to Chang Yü, used for the attack; the 革車 were heavier, and designed for purposes of defence. Li Ch'üan, it is true, says that the latter were light, but this seems hardly probable. Capt. Calthrop translates "chariots" and "supply wagons" respectively, but is not supported by any commentator. It is interesting to note the analogies between early Chinese warfare and that of the Homeric Greeks. In each case, the war-chariot was the important factor, forming as it did the nucleus round which was grouped a certain number of foot-soldiers. With regard to the numbers given here, we are informed that each swift chariot was accompanied by 75 footmen, and each heavy chariot by 25 footmen, so that the whole army would be divided up into a thousand battalions, each consisting of two chariots and a hundred men.

with provisions enough to carry them a thousand *li*,

2.78 modern *li* go to a mile. The length may have varied slightly since Sun Tzŭ's time.

the expenditure at home and at the front, including entertainment of guests, small items such as glue and paint,

2. 其用戰也勝久則鈍兵挫銳攻城則力屈

3. 久暴師則國用不足

and sums spent on chariots and armour, will reach the total of a thousand ounces of silver per day.

則, which follows 糧 in the *textus receptus*, is important as indicating the apodosis. In the text adopted by Capt. Calthrop it is omitted, so that he is led to give this meaningless translation of the opening sentence: "Now the requirements of War are such that we need 1,000 chariots," etc. The second 費, which is redundant, is omitted in the *Yü Lan*.

千金, like 千里 above, is meant to suggest a large but indefinite number. As the Chinese have never possessed gold coins, it is incorrect to translate it "1000 pieces of gold."

Such is the cost of raising an army of 100,000 men.

Capt. Calthrop adds: "You have the instruments of victory," which he seems to get from the first five characters of the next sentence.

2. When you engage in actual fighting, if victory is long in coming, the men's weapons will grow dull and their ardour will be damped.

The *Yü Lan* omits 勝; but though 勝久 is certainly a bold phrase, it is more likely to be right than not. Both in this place and in § 4, the *T'ung Tien* and *Yü Lan* read 頓 (in the sense of "to injure") instead of 鈍.

If you lay siege to a town, you will exhaust your strength.

As synonyms to 屈 are given 盡, 殫, 窮 and 困.

3. Again, if the campaign is protracted, the resources of the State will not be equal to the strain.

久暴師 means literally, "If there is long exposure of the army." Of 暴 in this sense K'ang Hsi cites an instance from the biography of 竇融 Tou Jung in the *Hou Han Shu*, where the commentary defines it by 露. Cf. also the following from the 戰國策: 將軍久暴露於外 "General, you have long been exposed to all weathers."

4. 夫鈍兵挫銳屈力殫貨則諸侯乘其弊而起
雖有智者不能善其後矣

5. 故兵聞拙速未睹巧之久也

4. Now, when your weapons are dulled, your ardour damped, your strength exhausted and your treasure spent, other chieftains will spring up to take advantage of your extremity. Then no man, however wise, will be able to avert the consequences that must ensue.

Following Tu Yu, I understand 善 in the sense of "to make good," i. e. to mend. But Tu Mu and Ho Shih explain it as "to make good plans" — for the future.

5. Thus, though we have heard of stupid haste in war, cleverness has never been seen associated with long delays.

This concise and difficult sentence is not well explained by any of the commentators. Ts'ao Kung, Li Ch'üan, Mêng Shih, Tu Yu, Tu Mu and Mei Yao-ch'ên have notes to the effect that a general, though naturally stupid, may nevertheless conquer through sheer force of rapidity. Ho Shih says: "Haste may be stupid, but at any rate it saves expenditure of energy and treasure; protracted operations may be very clever, but they bring calamity in their train." Wang Hsi evades the difficulty by remarking: "Lengthy operations mean an army growing old, wealth being expended, an empty exchequer and distress among the people; true cleverness insures against the occurrence of such calamities." Chang Yü says: "So long as victory can be attained, stupid haste is preferable to clever dilatoriness." Now Sun Tzŭ says nothing whatever, except possibly by implication, about ill-considered haste being better than ingenious but lengthy operations. What he does say is something much more guarded, namely that, while speed may sometimes be injudicious, tardiness can never be anything but foolish — if only because it means impoverishment to the nation. Capt. Calthrop indulges his imagination with the following: "Therefore it is acknowledged that war cannot be too short in duration. But though conducted with the utmost art, if long continuing, misfortunes do always appear." It is hardly worth while to note the total disappearance of 拙速 in this precious concoction. In considering the point raised here by Sun Tzŭ, the classic example of Fabius Cunctator will inevitably occur to the mind. That general deliberately measured the endurance of Rome against that of Hannibal's isolated army, because it seemed to him that the latter was more likely to suffer from a long campaign in a strange country. But it is quite a moot question whether his tactics would have proved successful in the long run. Their reversal, it is true, led to Cannae; but this only establishes a negative presumption in their favour.

6. 夫兵久而國利者未之有也
7. 故不盡知用兵之害者則不能盡知用兵之利也
8. 善用兵者役不再籍糧不三載
9. 取用於國因糧於敵故軍食可足也

6. There is no instance of a country having benefited from prolonged warfare.

The *Yü Lan* has 圖 instead of 國 — evidently the mistake of a scribe.

7. It is only one who is thoroughly acquainted with the evils of war that can thoroughly understand the profitable way of carrying it on.

That is, with rapidity. Only one who knows the disastrous effects of a long war can realise the supreme importance of rapidity in bringing it to a close. Only two commentators seem to favour this interpretation, but it fits well into the logic of the context, whereas the rendering, "He who does not know the evils of war cannot appreciate its benefits," is distinctly pointless.

8. The skilful soldier does not raise a second levy, neither are his supply-waggons loaded more than twice.

Once war is declared, he will not waste precious time in waiting for reinforcements, nor will he turn his army back for fresh supplies, but crosses the enemy's frontier without delay. This may seem an audacious policy to recommend, but with all great strategists, from Julius Caesar to Napoleon Buonaparte, the value of time — that is, being a little ahead of your opponent — has counted for more than either numerical superiority or the nicest calculations with regard to commissariat. 籍 is used in the sense of 賦. The *T'ung Tien* and *Yü Lan* have the inferior reading 藉. The commentators explain 不三載 by saying that the waggons are loaded once before passing the frontier, and that the army is met by a further consignment of supplies on the homeward march. The *Yü Lan*, however, reads 再 here as well.

9. Bring war material with you from home, but forage on the enemy. Thus the army will have food enough for its needs.

用, "things to be used," in the widest sense. It includes all the impedimenta of an army, apart from provisions.

10. 國之貧於師者遠輸遠輸則百姓貧
11. 近於師者貴賣貴賣則百姓財竭
12. 財竭則急於丘役

10. Poverty of the State exchequer causes an army to be maintained by contributions from a distance. Contributing to maintain an army at a distance causes the people to be impoverished.

The beginning of this sentence does not balance properly with the next, though obviously intended to do so. The arrangement, moreover, is so awkward that I cannot help suspecting some corruption in the text. It never seems to occur to Chinese commentators that an emendation may be necessary for the sense, and we get no help from them here. Sun Tzŭ says that the cause of the people's impoverishment is 遠輸; it is clear, therefore, that the words have reference to some system by which the husbandmen sent their contributions of corn to the army direct. But why should it fall on them to maintain an army in this way, except because the State or Government is too poor to do so? Assuming then that 貧 ought to stand first in the sentence in order to balance 近 (the fact that the two words rhyme is significant), and thus getting rid of 國之, we are still left with 於師, which latter word seems to me an obvious mistake for 國. "Poverty in the army" is an unlikely expression, especially as the general has just been warned not to encumber his army with a large quantity of supplies. If we suppose that 師 somehow got written here instead of 國 (a very simple supposition, as we have 近於師 in the next sentence), and that later on somebody, scenting a mistake, prefixed the gloss 國之 to 貧, without however erasing 於師, the whole muddle may be explained. My emended text then would be 貧於國者, etc.

11. On the other hand, the proximity of an army causes prices to go up; and high prices cause the people's substance to be drained away.

近, that is, as Wang Hsi says, before the army has left its own territory. Ts'ao Kung understands it of an army that has already crossed the frontier. Capt. Calthrop drops the 於, reading 近師者, but even so it is impossible to justify his translation "Repeated wars cause high prices."

12. When their substance is drained away, the peasantry will be afflicted by heavy exactions.

13. 力屈財殫中原內虛於家百姓之費十去其七

14. 公家之費破車罷馬甲冑矢弩戟楯蔽櫓丘牛大車十去其六

Cf. Mencius VII. 2. xiv. 2, where 丘民 has the same meaning as 丘役. 丘 was an ancient measure of land. The full table, as given by 司馬法, may not be out of place here: 6 尺＝1 步; 100 步＝1 畝; 100 畝＝1 夫; 3 夫＝1 屋; 3 屋＝1 井; 4 井＝1 邑; 4 邑＝1 丘; 4 丘＝1 甸. According to the *Chou Li*, there were nine husbandmen to a 井, which would assign to each man the goodly allowance of 100 畝 (of which 6.6 now go to an acre). What the values of these measures were in Sun Tzŭ's time is not known with any certainty. The lineal 尺, however, is supposed to have been about 20 cm. 急 may include levies of men, as well as other exactions.

13, 14. With this loss of substance and exhaustion of strength, the homes of the people will be stripped bare, and three-tenths of their incomes will be dissipated;

The *Yü Lan* omits 財殫. I would propose the emended reading 力屈則中, etc. In view of the fact that we have 財竭 in the two preceding paragraphs, it seems probable that 財 is a scribe's mistake for 則, 殫 having been added afterwards to make sense. 中原內虛於家, literally: "Within the middle plains there is emptiness in the homes." For 中原 cf. *Shih Ching* II. 3. VI. 3 and II. 5. II. 3. With regard to 十去其七, Tu Mu says: 家業十耗其七也, and Wang Hsi: 民費大半矣; that is, the people are mulcted not of $\frac{3}{10}$, but of $\frac{7}{10}$, of their income. But this is hardly to be extracted from our text. Ho Shih has a characteristic tag: 國以民為本民以食為天居人上者宜乎重惜 "The *people* being regarded as the essential part of the State, and *food* as the people's heaven, is it not right that those in authority should value and be careful of both?"

while Government expenses for broken chariots, worn-out horses, breast-plates and helmets, bows and arrows, spears and shields, protective mantlets, draught-oxen and heavy waggons, will amount to four-tenths of its total revenue.

15. 故智將務食於敵食敵一鍾當吾二十鍾萁
秆一石當吾二十石

16. 故殺敵者怒也取敵之利者貨也

17. 故車戰得車十乘已上賞其先得者而更其
旌旗車雜而乘之卒善而養之

The *Yü Lan* has several various readings here, the more important of which are 疲 for the less common 罷 (read *p'i²*), 干 for 蔽, and 兵牛 for 丘牛, which latter, if right, must mean "oxen from the country districts" (cf. *supra*, § 12). For the meaning of 櫓, see note on III, § 4. Capt. Calthrop omits to translate 丘牛大車.

15. Hence a wise general makes a point of foraging on the enemy. One cartload of the enemy's provisions is equivalent to twenty of one's own, and likewise a single picul of his provender is equivalent to twenty from one's own store.

Because twenty cartloads will be consumed in the process of transporting one cartload to the front. According to Ts'ao Kung, a 鍾 = 6 斛 4 斗, or 64 斗, but according to Mêng Shih, 10 斛 make a 鍾. The 石 picul consisted of 70 斤 catties (Tu Mu and others say 120). 萁秆, literally, "beanstalks and straw."

16. Now in order to kill the enemy, our men must be roused to anger; that there may be advantage from defeating the enemy, they must have their rewards.

These are two difficult sentences, which I have translated in accordance with Mei Yao-ch'ên's paraphrase. We may incontinently reject Capt. Calthrop's extraordinary translation of the first: "Wantonly to kill and destroy the enemy must be forbidden." Ts'ao Kung quotes a jingle current in his day: 軍無財士不來軍無賞士不往. Tu Mu says: "Rewards are necessary in order to make the soldiers see the advantage of beating the enemy; thus, when you capture spoils from the enemy, they must be used as rewards, so that all your men may have a keen desire to fight, each on his own account. Chang Yü takes 利 as the direct object of 取, which is not so good.

17. Therefore in chariot fighting, when ten or more chariots have been taken, those should be rewarded who took the first.

18. 是謂勝敵而益強
19. 故兵貴勝不貴久
20. 故知兵之將民之司命國家安危之主也

Capt. Calthrop's rendering is: "They who are the first to lay their hands on more than, ten of the enemy's chariots, should be encouraged." We should have expected the gallant captain to see that such Samson-like prowess deserved something more substantial than mere encouragement. T. omits 故, and has 以上 in place of the more archaic 已上.

Our own flags should be substituted for those of the enemy, and the chariots mingled and used in conjunction with ours. The captured soldiers should be kindly treated and kept.

18. This is called, using the conquered foe to augment one's own strength.

19. In war, then, let your great object be victory, not lengthy campaigns.

As Ho Shih remarks: 兵不可玩武不可黷 "War is not a thing to be trifled with." Sun Tzŭ here reiterates the main lesson which this chapter is intended to enforce.

20. Thus it may be known that the leader of armies is the arbiter of the people's fate, the man on whom it depends whether the nation shall be in peace or in peril.

In the original text, there is a 生 before the 民.

III. 謀攻篇

1. 孫子曰凡用兵之法全國爲上破國次之全
軍爲上破軍次之全旅爲上破旅次之全卒
爲上破卒次之全伍爲上破伍次之
2. 是故百戰百勝非善之善者也不戰而屈人
之兵善之善者也
3. 故上兵伐謀其次伐交其次伐兵下政攻城

III. ATTACK BY STRATAGEM.

1. Sun Tzǔ said: In the practical art of war, the best thing of all is to take the enemy's country whole and intact; to shatter and destroy it is not so good. So, too, it is better to capture an army entire than to destroy it, to capture a regiment, a detachment or a company entire than to destroy them.

A 軍 "army corps," according to Ssǔ-ma Fa, consisted nominally of 12500 men; according to Ts'ao Kung, a 旅 contained 500 men, a 卒 any number between 100 and 500, and a 伍 any number between 5 and 100. For the last two, however, Chang Yü gives the exact figures of 100 and 5 respectively.

2. Hence to fight and conquer in all your battles is not supreme excellence; supreme excellence consists in breaking the enemy's resistance without fighting.

Here again, no modern strategist but will approve the words of the old Chinese general. Moltke's greatest triumph, the capitulation of the huge French army at Sedan, was won practically without bloodshed.

3. Thus the highest form of generalship is to baulk the enemy's plans;

4. 攻城之法爲不得已修櫓轒轀具器械三月
而後成距闉又三月而後已

I. e., as Li Ch'üan says (伐其始謀也), in their very inception.
Perhaps the word "baulk" falls short of expressing the full force of 伐,
which implies not an attitude of defence, whereby one might be content
to foil the enemy's stratagems one after another, but an active policy of
counter-attack. Ho Shih puts this very clearly in his note: "When the
enemy has made a plan of attack against us, we must anticipate him by
delivering our own attack first."

the next best is to prevent the junction of the enemy's
forces;

Isolating him from his allies. We must not forget that Sun Tzŭ, in
speaking of hostilities, always has in mind the numerous states or principalities into which the China of his day was split up.

the next in order is to attack the enemy's army in the field;

When he is already in full strength.

and the worst policy of all is to besiege walled cities.

The use of the word 政 is somewhat unusual, which may account for
the reading of the modern text: 其下攻城.

4. The rule is, not to besiege walled cities if it can
possibly be avoided.

Another sound piece of military theory. Had the Boers acted upon it
in 1899, and refrained from dissipating their strength before Kimberley,
Mafeking, or even Ladysmith, it is more than probable that they would
have been masters of the situation before the British were ready seriously
to oppose them.

The preparation of mantlets, movable shelters, and various
implements of war, will take up three whole months;

It is not quite clear what 櫓 were. Ts'ao Kung simply defines them
as 大楯 "large shields," but we get a better idea of them from Li Ch'üan,
who says they were to protect the heads of those who were assaulting
the city walls at close quarters. This seems to suggest a sort of Roman
testudo, ready made. Tu Mu says they were "what are now termed
彭排" (wheeled vehicles used in repelling attacks, according to K'ang
Hsi), but this is denied by Ch'ên Hao. See *supra*, II. 14. The name
is also applied to turrets on city walls. Of 轒轀 (*fên yün*) we get

5. 將不勝其忿而蟻附之殺士三分之一而城
不拔者此攻之災

6. 故善用兵者屈人之兵而非戰也拔人之城
而非攻也毀人之國而非久也

a fairly clear description from several commentators. They were wooden missile-proof structures on four wheels, propelled from within, covered over with raw hides, and used in sieges to convey parties of men to and from the walls, for the purpose of filling up the encircling moat with earth. Tu Mu adds that they are now called 木驢 "wooden donkeys." Capt. Calthrop wrongly translates the term, "battering-rams." I follow Ts'ao Kung in taking 具 as a verb, co-ordinate and synonymous with 修. Those commentators who regard 修 as an adjective equivalent to 長 "long," make 具 presumably into a noun.

and the piling up of mounds over against the walls will take three months more.

The 距闉 (or 堙, in the modern text) were great mounds or ramparts of earth heaped up to the level of the enemy's walls in order to discover the weak points in the defence, and also to destroy the 樓櫓 fortified turrets mentioned in the preceding note. Tu Yu quotes the Tso Chuan: 楚司馬子反乘堙而窺宋城也.

5. The general, unable to control his irritation, will launch his men to the assault like swarming ants,

Capt. Calthrop unaccountably omits this vivid simile, which, as Ts'ao Kung says, is taken from the spectacle of an army of ants climbing a wall. The meaning is that the general, losing patience at the long delay, may make a premature attempt to storm the place before his engines of war are ready.

with the result that one-third of his men are slain, while the town still remains untaken. Such are the disastrous effects of a siege.

We are reminded of the terrible losses of the Japanese before Port Arthur, in the most recent siege which history has to record. The *T'ung Tien* reads 不勝心之忿…則殺士卒…攻城之災. For 其忿 the *Yü Lan* has 心怒. Capt. Calthrop does not translate 而城不拔者, and mistranslates 此攻之災.

6. Therefore the skilful leader subdues the enemy's troops without any fighting; he captures their cities with-

7. 必以全爭於天下故兵不頓而利可全此謀
攻之法也.

8. 故用兵之法十則圍之五則攻之倍則分之

out laying siege to them; he overthrows their kingdom without lengthy operations in the field.

Chia Lin notes that he only overthrows the 國 , that is, the Government, but does no harm to individuals. The classical instance is Wu Wang, who after having put an end to the Yin dynasty was acclaimed "Father and mother of the people."

7. With his forces intact he will dispute the mastery of the Empire, and thus, without losing a man, his triumph will be complete.

Owing to the double meanings of 兵, 頓 [= 鈍] and 利, the latter part of the sentence is susceptible of quite a different meaning: "And thus, the weapon not being blunted by use, its keenness remains perfect." Chang Yü says that 利 is "the advantage of a prosperous kingdom and a strong army."

This is the method of attacking by stratagem.

8. It is the rule in war, if our forces are ten to the enemy's one, to surround him; if five to one, to attack him;

Straightaway, without waiting for any further advantage.

if twice as numerous, to divide our army into two.

Note that 之 does not refer to the enemy, as in the two preceding clauses. This sudden change of object is quite common in Chinese. Tu Mu takes exception to the saying; and at first sight, indeed, it appears to violate a fundamental principle of war. Ts'ao Kung, however, gives a clue to Sun Tzŭ's meaning: 以二敵一則一術爲正一 術爲奇 "Being two to the enemy's one, we may use one part of our army in the regular way, and the other for some special diversion." [For explanation of 正 and 奇, see V. 3, note.] Chang Yü thus further elucidates the point: "If our force is twice as numerous as that of the enemy, it should be split up into two divisions, one to meet the enemy in front, and one to fall upon his rear; if he replies to the frontal attack, he may be crushed from behind; if to the rearward attack, he may be crushed in front. This is what is meant by saying that "one part may be used in the regular way, and the other for some special diversion." Tu Mu does not understand that dividing one's army is simply an irregular, just as concentrating it is the regular, strategical method, and he is too hasty in calling this a mistake."

9. 敵則能戰之少則能逃之不若則能避之

10. 故小敵之堅大敵之擒也

11. 夫將者國之輔也輔周則國必強輔隙則國
必弱

12. 故君之所以患於軍者三

9. If equally matched, we can offer battle;

Li Ch'üan, followed by Ho Shih, gives the following paraphrase:
主客力敵惟善者戰 "If attackers and attacked are equally
matched in strength, only the able general will fight." He thus takes
能 as though it were 能者, which is awkward.

if slightly inferior in numbers, we can avoid the enemy;

The *Tʻu Shu* has 守 instead of 逃, which is hardly distinguishable in sense
from 避 in the next clause. The meaning, "we can *watch* the enemy," is
certainly a great improvement on the above; but unfortunately there appears to be no very good authority for the variant. Chang Yü reminds
us that the saying only applies if the other factors are equal; a small
difference in numbers is often more than counterbalanced by superior
energy and discipline.

if quite unequal in every way, we can flee from him.

10. Hence, though an obstinate fight may be made
by a small force, in the end it must be captured by the
larger force.

In other words: "C'est magnifique; mais ce n'est pas la guerre."

11. Now the general is the bulwark of the State: if
the bulwark is complete at all points, the State will be
strong; if the bulwark is defective, the State will be weak.

隙 cannot be restricted to anything so particular as in Capt. Calthrop's
translation, "divided in his allegiance." It is simply keeping up the
metaphor suggested by 周. As Li Ch'üan tersely puts it: 隙缺也
將才不備兵必弱 "*Chʻi*, gap, indicates deficiency; if the
general's ability is not perfect (i. e. if he is not thoroughly versed in his
profession), his army will lack strength."

12. There are three ways in which a ruler can bring
misfortune upon his army: —

13. 不知軍之不可以進而謂之進不知軍之不
　　可以退而謂之退是謂縻軍
14. 不知三軍之事而同三軍之政者則軍士惑
　　矣

13. (1) By commanding the army to advance or to retreat, being ignorant of the fact that it cannot obey. This is called hobbling the army.

Ts'ao Kung weakly defines 縻 as 御 "control," "direct." Cf. § 17 *ad fin.* But in reality it is one of those graphic metaphors which from time to time illuminate Sun Tzŭ's work, and is rightly explained by Li Ch'üan as = 絆. He adds the comment: 如絆驥足無馳驟也. "It is like tying together the legs of a thoroughbred, so that it is unable to gallop." One would naturally think of "the ruler" in this passage as being at home, and trying to direct the movements of his army from a distance. But the commentators understand just the reverse, and quote the saying of T'ai Kung: 國不可以從外治軍不可以從中御 "A kingdom should not be governed from without, an army should not be directed from within." Of course it is true that, during an engagement, or when in close touch with the enemy, the general should not be in the thick of his own troops, but a little distance apart. Otherwise, he will be liable to misjudge the position as a whole, and give wrong orders.

14. (2) By attempting to govern an army in the same way as he administers a kingdom, being ignorant of the conditions which obtain in an army. This causes restlessness in the soldier's minds.

Ts'ao Kung's note is: 軍容不入國國容不入軍禮不可以治兵也, which may be freely translated: "The military sphere and the civil sphere are wholly distinct; you can't handle an army in kid gloves." And Chang Yü says: "Humanity and justice (仁義) are the principles on which to govern a state, but not an army; opportunism and flexibility (權變), on the other hand, are military rather than civic virtues." 同三軍之政, "to assimilate the governing of an army" — to that of a State, understood. The *T'ung Tien* has 欲 inserted before 同, here and in § 15.

15. 不知三軍之權而同三軍之任則軍士疑矣

16. 三軍既惑且疑則諸侯之難至矣是謂亂軍引勝

17. 故知勝有五知可以戰與不可以戰者勝識眾寡之用者勝上下同欲者勝以虞待不虞者勝將能而君不御者勝此五者知勝之道也

15. (3) By employing the officers of his army without discrimination,

That is, he is not careful to use the right man in the right place.

through ignorance of the military principle of adaptation to circumstances.　This shakes the confidence of the soldiers.

I follow Mei Yao-ch'ên here.　The other commentators make 不知 etc. refer, not to the ruler, as in §§ 13, 14, but to the officers he employs. Thus Tu Yu says: 將若不知權變不可付以勢位 "If a general is ignorant of the principle of adaptability, he must not be entrusted with a position of authority."　Tu Mu quotes 黃石公: "The skilful employer of men will employ the wise man, the brave man, the covetous man, and the stupid man.　For the wise man delights in establishing his merit, the brave man likes to show his courage in action, the covetous man is quick at seizing advantages, and the stupid man has no fear of death."　The T'ung Tien reads 軍覆疑, which Tu Yu explains as 覆敗 "is utterly defeated."　Capt. Calthrop gives a very inaccurate rendering: "Ignorant of the situation of the army, to interfere in its dispositions."

16. But when the army is restless and distrustful, trouble is sure to come from the other feudal princes.　This is simply bringing anarchy into the army, and flinging victory away.

Most of the commentators take 引 in the sense of 奪, which it seems to bear also in the Li Chi, 玉藻, I. 18. [卻 is there given as its equivalent, but Legge tries notwithstanding to retain the more usual sense, translating "draw...back," which is hardly defensible.] Tu Mu and Wang Hsi, however, think 引勝 means "leading up to the enemy's victory."

17. Thus we may know that there are five essentials

18. 故曰知彼知己百戰不殆不知彼而知己一勝一負不知彼不知己每戰必殆

for victory: (1) He will win who knows when to fight and when not to fight.

Chang Yü says: "If he can fight, he advances and takes the offensive; if he cannot fight, he retreats and remains on the defensive. He will invariably conquer who knows whether it is right to take the offensive or the defensive."

(2) He will win who knows how to handle both superior and inferior forces.

This is not merely the general's ability to estimate numbers correctly, as Li Ch'üan and others make out. Chang Yü expounds the saying more satisfactorily: "By applying the art of war, it is possible with a lesser force to defeat a greater, and *vice versá*. The secret lies in an eye for locality, and in not letting the right moment slip. Thus Wu Tzǔ says: 'With a superior force, make for easy ground; with an inferior one, make for difficult ground.'"

(3) He will win whose army is animated by the same spirit throughout all its ranks.

Ts'ao Kung refers 上 下 less well to sovereign and subjects.

(4) He will win who, prepared himself, waits to take the enemy unprepared.

(5) He will win who has military capacity and is not interfered with by the sovereign.

Tu Yu quotes 王子 as saying: 指授在君決戰在將也 "It is the sovereign's function to give broad instructions, but to decide on battle is the function of the general." It is needless to dilate on the military disasters which have been caused by undue interference with operations in the field on the part of the home government. Napoleon undoubtedly owed much of his extraordinary success to the fact that he was not hampered by any central authority, — that he was, in fact, 將 and 君 in one.

Victory lies in the knowledge of these five points.

Literally, "These five things are knowledge of the principle of victory."

18. Hence the saying: If you know the enemy and know yourself, you need not fear the result of a hundred

battles. If you know yourself but not the enemy, for every victory gained you will also suffer a defeat.

Li Ch'üan cites the case of 苻 堅 Fu Chien, prince of 秦 Ch'in, who in 383 A.D. marched with a vast army against the 晉 Chin Emperor. When warned not to despise an enemy who could command the services of such men as 謝 安 Hsieh An and 桓 沖 Huan Ch'ung, he boastfully replied: "I have the population of eight provinces at my back, infantry and horsemen to the number of one million; why, they could dam up the Yangtsze River itself by merely throwing their whips into the stream. What danger have I to fear?" Nevertheless, his forces were soon after disastrously routed at the 淝 Fei River, and he was obliged to beat a hasty retreat.

If you know neither the enemy nor yourself, you will succumb in every battle.

The modern text, represented by the 北 堂 書 鈔 and *T'u Shu*, has 必 敗, which I should be inclined to adopt in preference to 殆 here, though the *T'ung Tien* and *Yü Lan* both have the latter. Chang Yü offers the best commentary on 知 彼 知 己. He says that these words "have reference to attack and defence: knowing the enemy enables you to take the offensive, knowing yourself enables you to stand on the defensive." He adds: 攻 是 守 之 機 守 是 攻 之 策 "Attack is the secret of defence; defence is the planning of an attack." It would be hard to find a better epitome of the root-principle of war.

IV. 形篇

1. 孫子曰昔之善戰者先爲不可勝以待敵之
可勝
2. 不可勝在己可勝在敵

IV. TACTICAL DISPOSITIONS.

形 is a very comprehensive and somewhat vague term. Literally, "form," "body," it comes to mean "appearance," "attitude" or "disposition;" and here it is best taken as something between, or perhaps combining, "tactics" and "disposition of troops." Ts'ao Kung explains it as 軍之形也、我動彼應兩敵相察情也 "marching and counter-marching on the part of the two armies with a view to discovering each other's condition." Tu Mu says: "It is through the 形 dispositions of an army that its condition may be discovered. Conceal your dispositions (無形), and your condition will remain secret, which leads to victory; show your dispositions, and your condition will become patent, which leads to defeat." Wang Hsi remarks that the good general can 變化其形因敵以制勝 "secure success by modifying his tactics to meet those of the enemy." In the modern text, the title of the chapter appears as 軍形, which Capt. Calthrop incorrectly translates "the order of battle."

1. Sun Tzŭ said: The good fighters of old first put themselves beyond the possibility of defeat, and then waited for an opportunity of defeating the enemy.

2. To secure ourselves against defeat lies in our own hands, but the opportunity of defeating the enemy is provided by the enemy himself.

That is, of course, by a mistake on his part. Capt. Calthrop has: "The causes of defeat come from within; victory is born in the enemy's camp," which, though certainly an improvement on his previous attempt, is still incorrect.

3. 故善戰者能爲不可勝不能使敵必可勝
4. 故曰勝可知而不可爲
5. 不可勝者守也可勝者攻也
6. 守則不足攻則有餘
7. 善守者藏於九地之下善攻者動於九天之
上故能自保而全勝也

3. Thus the good fighter is able to secure himself against defeat,

"By concealing the disposition of his troops, covering up his tracks, and taking unremitting precautions" (Chang Yü).

but cannot make certain of defeating the enemy.

The original text reads 使敵之可勝, which the modern text has further modified into 使敵之必可勝. Capt. Calthrop makes out the impossible meaning, "and further render the enemy incapable of victory."

4. Hence the saying: One may *know* how to conquer without being able to *do* it.

Capt. Calthrop translates: "The conditions necessary for victory may be present, but they cannot always be obtained," which is more or less unintelligible.

5. Security against defeat implies defensive tactics; ability to defeat the enemy means taking the offensive.

For 不可勝 I retain the sense which it undoubtedly bears in §§ 1—3, in spite of the fact that the commentators are all against me. The meaning they give, "He who cannot conquer takes the defensive," is plausible enough, but it is highly improbable that 勝 should suddenly become active in this way. An incorrect variant in the *Yü Lan* is 不可勝則守可勝則攻.

6. Standing on the defensive indicates insufficient strength; attacking, a superabundance of strength.

7. The general who is skilled in defence hides in the most secret recesses of the earth;

Literally, "hides under the ninth earth," which is a metaphor indicating the utmost secrecy and concealment, so that the enemy may not know

8. 見勝不過衆人之所知非善之善者也

9. 戰勝而天下曰善非善之善者也

his whereabouts. The 九地 of this passage have of course no connection with the 九地 "Nine situations" of chap. XI.

he who is skilled in attack flashes forth from the topmost heights of heaven.

Another metaphor, implying that he falls on his adversary like a thunderbolt, against which there is no time to prepare. This is the opinion of most of the commentators, though Ts'ao Kung, followed by Tu Yu, explains 地 as the hills, rivers, and other natural features which will afford shelter or protection to the attacked, and 天 as the phases of weather which may be turned to account by the attacking party. Capt. Calthrop's "The skilful in attack push to the topmost heaven" conveys no meaning at all.

Thus on the one hand we have ability to protect ourselves; on the other, a victory that is complete.

Capt. Calthrop draws on a fertile imagination for the following: "If these precepts be observed, victory is certain."

8. To see victory only when it is within the ken of the common herd is not the acme of excellence.

As Ts'ao Kung remarks, 當見未萌 "the thing is to see the plant before it has germinated," to foresee the event before the action has begun. Li Ch'üan alludes to the story of Han Hsin who, when about to attack the vastly superior army of 趙 Chao, which was strongly entrenched in the city of 成安 Ch'êng-an, said to his officers: "Gentlemen, we are going to annihilate the enemy, and shall meet again at dinner." The officers hardly took his words seriously, and gave a very dubious assent. But Han Hsin had already worked out in his mind the details of a clever stratagem, whereby, as he foresaw, he was able to capture the city and inflict a crushing defeat on his adversary. For the full story, see 前漢書, chap. 34, 韓信傳. Capt. Calthrop again blunders badly with: "A victory, even if popularly proclaimed as such by the common folk, may not be a true success."

9. Neither is it the acme of excellence if you fight and conquer and the whole Empire says, "Well done!"

True excellence being, as Tu Mu says: 陰謀潛運攻心伐謀勝敵之日曾不血刃 "To plan secretly, to move sur-

10. 故舉秋毫不爲多力見日月不爲明目聞雷
霆不爲聰耳

11. 古之所謂善戰者勝勝易勝者也

12. 故善戰者之勝也無智名無勇功

reptitiously, to foil the enemy's intentions and baulk his schemes, so that at last the day may be won without shedding a drop of blood." Sun Tzŭ reserves his approbation for things that

> "the world's coarse thumb
> And finger fail to plumb."

10. To lift an autumn hair is no sign of great strength;

秋毫 is explained as the fur of a hare, which is finest in autumn, when it begins to grow afresh. The phrase is a very common one in Chinese writers. Cf. Mencius, I. 1. vii. 10, and Chuang Tzŭ, 知北遊, *et al.*

to see sun and moon is no sign of sharp sight; to hear the noise of thunder is no sign of a quick ear.

Ho Shih gives as real instances of strength, sharp sight and quick hearing: 烏獲 Wu Huo, who could lift a tripod weighing 250 stone; 離朱 Li Chu, who at a distance of a hundred paces could see objects no bigger than a mustard seed; and 師曠 Shih K'uang, a blind musician who could hear the footsteps of a mosquito.

11. What the ancients called a clever fighter is one who not only wins, but excels in winning with ease.

The original text, followed by the *T'u Shu*, has 勝於易勝者也. But this is an alteration evidently intended to smooth the awkwardness of 勝勝易勝者也, which means literally: "one who, conquering, excels in easy conquering." Mei Yao-ch'ên says: "He who only sees the obvious, wins his battles with difficulty; he who looks below the surface of things, wins with ease."

12. Hence his victories bring him neither reputation for wisdom nor credit for courage.

Tu Mu explains this very well: "Inasmuch as his victories are gained over circumstances that have not come to light, the world at large knows nothing of them, and he wins no reputation for wisdom; inasmuch as the hostile state submits before there has been any bloodshed, he receives no credit for courage."

13. 故其戰勝不忒不忒者其所措必勝勝已敗
者也

14. 故善戰者立於不敗之地而不失敵之敗也

15. 是故勝兵先勝而後求戰敗兵先戰而後求
勝

13. He wins his battles by making no mistakes.

Ch'ên Hao says: "He plans no superfluous marches, he devises no futile attacks." The connection of ideas is thus explained by Chang Yü: "One who seeks to conquer by sheer strength, clever though he may be at winning pitched battles, is also liable on occasion to be vanquished; whereas he who can look into the future and discern conditions that are not yet manifest, will never make a blunder and therefore invariably win." Li Ch'üan thinks that the character 忒 should be 貳 "to have doubts." But it is better not to tamper with the text, especially when no improvement in sense is the result.

Making no mistakes is what establishes the certainty of victory, for it means conquering an enemy that is already defeated.

The *T'u Shu* omits 必. 措 is here = 置. Chia Lin says it is put for 錯 in the sense of 雜; but this is far-fetched. Capt. Calthrop altogether ignores the important word 忒.

14. Hence the skilful fighter puts himself into a position which makes defeat impossible, and does not miss the moment for defeating the enemy.

A 不可爲之計 "counsel of perfection," as Tu Mu truly observes. 地 need not be confined strictly to the actual ground occupied by the troops. It includes all the arrangements and preparations which a wise general will make to increase the safety of his army.

15. Thus it is that in war the victorious strategist only seeks battle after the victory has been won, whereas he who is destined to defeat first fights and afterwards looks for victory.

Ho Shih thus expounds the paradox: "In warfare, first lay plans which will ensure victory, and then lead your army to battle; if you will not begin with stratagem but rely on brute strength alone, victory will no longer be assured."

16. 善用兵者修道而保法故能爲勝敗之政
17. 兵法一曰度二曰量三曰數四曰稱五曰勝
18. 地生度度生量量生數數生稱稱生勝

16. The consummate leader cultivates the moral law, and strictly adheres to method and discipline;

For 道 and 法, see *supra*, I. 4 sqq. I think that Chang Yü is wrong in altering their signification here, and taking them as 爲戰之道 and 制敵之法 respectively.

thus it is in his power to control success.

17. In respect of military method, we have, firstly, Measurement; secondly, Estimation of quantity; thirdly, Calculation; fourthly, Balancing of chances; fifthly, Victory.

18. Measurement owes its existence to Earth; Estimation of quantity to Measurement; Calculation to Estimation of quantity; Balancing of chances to Calculation; and Victory to Balancing of chances.

It is not easy to distinguish the four terms 度量數稱 very clearly. The first seems to be surveying and measurement of the ground, which enable us to 量 form an estimate of the enemy's strength, and to 數 make calculations based on the data thus obtained; we are thus led to 稱 a general weighing-up, or comparison of the enemy's chances with our own; if the latter turn the scale, then 勝 victory ensues. The chief difficulty lies in 數, which some commentators take as a calculation of *numbers*, thereby making it nearly synonymous with 量. Perhaps 量 is rather a consideration of the enemy's general position or condition (情 or 形勢), while 數 is the estimate of his numerical strength. On the other hand, Tu Mu defines 數 as 機數, and adds: 強弱已定然後能用機變數也 "the question of relative strength having been settled, we can bring the varied resources of cunning into play." Ho Shih seconds this interpretation, which is weakened, however, by the fact that 稱 is given as logically consequent on 數; this certainly points to the latter being a calculation of numbers. Of Capt. Calthrop's version the less said the better.

19. 故勝兵若以鎰稱銖敗兵若以銖稱鎰

20. 勝者之戰民也若決積水於千仞之谿者形
也

19. A victorious army opposed to a routed one, is as a pound's weight placed in the scale against a single grain.

Literally, "a victorious army is like an 鎰 *i* (20 oz.) weighed against a 銖 *shu* ($\frac{1}{24}$ oz.); a routed army as a *shu* weighed against an *i*." The point is simply the enormous advantage which a disciplined force, flushed with victory, has over one demoralised by defeat. Legge, in his note on Mencius, I. 2. ix. 2, makes the 鎰 to be 24 Chinese ounces, and corrects Chu Hsi's statement that it equalled 20 oz. only. But Li Ch'üan of the T'ang dynasty here gives the same figure as Chu Hsi.

20. The onrush of a conquering force is like the bursting of pent-up waters into a chasm a thousand fathoms deep. So much for tactical dispositions.

The construction here is slightly awkward and elliptical, but the general sense is plain. The *T'u Shu* omits 民也. A 仞 = 8 尺 or Chinese feet.

V. 埶篇

1. 孫子曰凡治眾如治寡分數是也
2. 鬥眾如鬥寡形名是也

V. ENERGY.

埶 here is said to be an older form of 勢; Sun Tzŭ, however, would seem to have used the former in the sense of "power," and the latter only in the sense of "circumstances." The fuller title 兵勢 is found in the *T'u Shu* and the modern text. Wang Hsi expands it into 積勢之變 "the application, in various ways, of accumulated power;" and Chang Yü says: 兵勢以成然後任勢以取勝 "When the soldiers' energy has reached its height, it may be used to secure victory."

1. Sun Tzŭ said: The control of a large force is the same in principle as the control of a few men: it is merely a question of dividing up their numbers.

That is, cutting up the army into regiments, companies, etc., with subordinate officers in command of each. Tu Mu reminds us of Han Hsin's famous reply to the first Han Emperor, who once said to him: "How large an army do you think I could lead?" "Not more than 100,000 men, your Majesty." "And you?" asked the Emperor. "Oh!" he answered, "the more the better" (多多益辦耳). Chang Yü gives the following curious table of the subdivisions of an army: — 5 men make a 列; 2 列 make a 火; 5 火 make a 隊; 2 隊 make a 官; 2 官 make a 曲; 2 曲 make a 部; 2 部 make a 校; 2 校 make a 裨; 2 裨 make a 軍. A 軍 or army corps thus works out at 3200 men. But cf. III. § 1, note. For 曲, see I. § 10. It is possible that 官 in that paragraph may also be used in the above technical sense.

2. Fighting with a large army under your command is nowise different from fighting with a small one: it is merely a question of instituting signs and signals.

3

3. 三軍之眾可使必受敵而無敗者奇正是也

One must be careful to avoid translating 鬥眾 "fighting *against* a large number," no reference to the enemy being intended. 形 is explained by Ts'ao Kung as denoting flags and banners, by means of which every soldier may recognise his own particular regiment or company, and thus confusion may be prevented. 名 he explains as drums and gongs, which from the earliest times were used to sound the advance and the retreat respectively. Tu Mu defines 形 as 陳形 "marshalling the troops in order," and takes 名 as the flags and banners. Wang Hsi also dissents from Ts'ao Kung, referring 形 to the ordering of the troops by means of banners, drums and gongs, and 名 to the various names by which the regiments might be distinguished. There is much to be said for this view.

3. To ensure that your whole host may withstand the brunt of the enemy's attack and remain unshaken — this is effected by manœuvres direct and indirect.

For 必, there is another reading 畢, "all together," adopted by Wang Hsi and Chang Yü. We now come to one of the most interesting parts of Sun Tzŭ's treatise, the discussion of the 正 and the 奇. As it is by no means easy to grasp the full significance of these two terms, or to render them at all consistently by good English equivalents, it may be as well to tabulate some of the commentators' remarks on the subject before proceeding further. Li Ch'üan: 當敵爲正傍出爲奇 "Facing the enemy is *chêng*, making lateral diversions is *ch'i*." Chia Lin: 當敵以正陳取勝以奇兵 "In presence of the enemy, your troops should be arrayed in normal fashion, but in order to secure victory abnormal manœuvres must be employed." Mei Yao-ch'ên: 動爲奇靜爲正靜以待之動以勝之 "*Ch'i* is active, *chêng* is passive; passivity means waiting for an opportunity, activity brings the victory itself." Ho Shih: 我之正使敵視之爲奇我之奇使敵視之爲正正亦爲奇奇亦爲正 "We must cause the enemy to regard our straightforward attack as one that is secretly designed, and *vice versâ;* thus *chêng* may also be *ch'i*, and *ch'i* may also be *chêng*." He instances the famous exploit of Han Hsin, who when marching ostensibly against 臨晉 Lin-chin (now 朝邑 Chao-i in Shensi), suddenly threw a large force across the Yellow River in wooden tubs, utterly disconcerting his opponent. [*Ch'ien Han Shu*, ch. 34.] Here, we are told, the march on Lin-chin was 正, and the surprise manœuvre was 奇. Chang Yü gives the following summary of opinions on the words: "Military writers

4. 兵之所加如以碬投卵者虛實是也
5. 凡戰者以正合以奇勝

do not all agree with regard to the meaning of *ch'i* and *chêng*. 尉繚子 Wei Liao Tzŭ [4ᵗʰ cent. B.C.] says: 正兵貴先奇兵貴後 'Direct warfare favours frontal attacks, indirect warfare attacks from the rear.' Ts'ao Kung says: 'Going straight out to join battle is a direct operation; appearing on the enemy's rear is an indirect manœuvre.' 李衛公 Li Wei-kung [6ᵗʰ and 7ᵗʰ cent. A.D.] says: 'In war, to march straight ahead is *chêng;* turning movements, on the other hand, are *ch'i.'* These writers simply regard *chêng* as *chêng,* and *ch'i* as *ch'i;* they do not note that the two are mutually interchangeable and run into each other like the two sides of a circle [see *infra,* § 11]. A comment of the T'ang Emperor T'ai Tsung goes to the root of the matter: 'A *ch'i* manœuvre may be *chêng,* if we make the enemy look upon it as *chêng;* then our real attack will be *ch'i,* and *vice versâ.* The whole secret lies in confusing the enemy, so that he cannot fathom our real intent.'" To put it perhaps a little more clearly: any attack or other operation is 正, on which the enemy has had his attention fixed; whereas that is 奇, which takes him by surprise or comes from an unexpected quarter. If the enemy perceives a movement which is meant to be 奇, it immediately becomes 正.

4. That the impact of your army may be like a grind-stone dashed against an egg — this is effected by the science of weak points and strong.

虛實, literally "the hollow and the solid," is the title of chap. VI. 碬 *tuan* is the *T'u Shu* reading, 碬 *hsia* that of the standard text. It appears from K'ang Hsi that there has been much confusion between the two characters, and indeed, it is probable that one of them has really crept into the language as a mistake for the other.

5. In all fighting, the direct method may be used for joining battle, but indirect methods will be needed in order to secure victory.

Chang Yü says: 徐發奇兵或擣其旁或擊其後 "Steadily develop indirect tactics, either by pounding the enemy's flanks or falling on his rear." A brilliant example of "indirect tactics" which decided the fortunes of a campaign was Lord Roberts' night march round the Peiwar Kotal in the second Afghan war.*

---

* "Forty-one Years in India," chap. 46.

6. 故善出奇者無窮如天地不竭如江河終而
復始日月是也死而復生四時是也

7. 聲不過五五聲之變不可勝聽也

8. 色不過五五色之變不可勝觀也

9. 味不過五五味之變不可勝嘗也

6. Indirect tactics, efficiently applied, are inexhaustible as Heaven and Earth, unending as the flow of rivers and streams;

奇 is the universally accepted emendation for 兵, the reading of the 北堂書鈔.

like the sun and moon, they end but to begin anew; like the four seasons, they pass away but to return once more.

Tu Yu and Chang Yü understand this of the permutations of 奇 and 正. But at present Sun Tzŭ is not speaking of 正 at all, unless, indeed, we suppose with 鄭友賢 Chêng Yu-hsien that a clause relating to it has fallen out of the text. Of course, as has already been pointed out, the two are so inextricably interwoven in all military operations, that they cannot really be considered apart. Here we simply have an expression, in figurative language, of the almost infinite resource of a great leader.

7. There are not more than five musical notes,

宮商角徵羽.

yet the combinations of these five give rise to more melodies than can ever be heard.

8. There are not more than five primary colours,

青黃赤白黑 blue, yellow, red, white and black.

yet in combination they produce more hues than can ever be seen.

9. There are not more than five cardinal tastes,

酸辛鹹甘苦 sour, acrid, salt, sweet, bitter.

yet combinations of them yield more flavours than can ever be tasted.

10. 戰埶不過奇正奇正之變不可勝窮也
11. 奇正相生如循環之無端孰能窮之
12. 激水之疾至於漂石者埶也
13. 鷙鳥之疾至於毀折者節也

10. In battle, there are not more than two methods of attack — the direct and the indirect; yet these two in combination give rise to an endless series of manœuvres.

11. The direct and the indirect lead on to each other in turn. It is like moving in a circle — you never come to an end. Who can exhaust the possibilities of their combination?

The *T'u Shu* adds 哉. The final 之 may refer either to the circle or, more probably, to the 奇正之變 understood. Capt. Calthrop is wrong with: "They are a mystery that none can penetrate."

12. The onset of troops is like the rush of a torrent which will even roll stones along in its course.

13. The quality of decision is like the well-timed swoop of a falcon which enables it to strike and destroy its victim.

For 疾 the *Yü Lan* reads 擊, which is also supported by a quotation in the 呂氏春秋 [3rd cent. B.C.]. 節 in this context is a word which really defies the best efforts of the translator. Tu Mu says that it is equivalent to 節量遠近 "the measurement or estimation of distance." But this meaning does not quite fit the illustrative simile in § 15. As applied to the falcon, it seems to me to denote that instinct of *self-restraint* which keeps the bird from swooping on its quarry until the right moment, together with the power of judging when the right moment has arrived. The analogous quality in soldiers is the highly important one of being able to reserve their fire until the very instant at which it will be most effective. When the "Victory" went into action at Trafalgar at hardly more than drifting pace, she was for several minutes exposed to a storm of shot and shell before replying with a single gun. Nelson coolly waited until he was within close range, when the broadside he brought to bear worked fearful havoc on the enemy's nearest ships. That was a case of 節.

14. 是故善戰者其埶險其節短
15. 埶如彍弩節如發機
16. 紛紛紜紜鬥亂而不可亂也渾渾沌沌形圓
而不可敗也

14. Therefore the good fighter will be terrible in his onset, and prompt in his decision.

Tu Yu defines 節 here by the word 斷, which is very like "decision" in English. 短 is certainly used in a very unusual sense, even if, as the commentators say, it = 近. This would have reference to the measurement of distance mentioned above, letting the enemy get near before striking. But I cannot help thinking that Sun Tzǔ meant to use the word in a figurative sense comparable to our own idiom "short and sharp." Cf. Wang Hsi's note, which after describing the falcon's mode of attack, proceeds: 兵之乘機當如是耳 "This is just how the 'psychological moment' should be seized in war." I do not care for Capt. Calthrop's rendering: "The spirit of the good fighter is terrifying, his occasions sudden."

15. Energy may be likened to the bending of a cross-bow; decision, to the releasing of the trigger.

"Energy" seems to be the best equivalent here for 埶, because the comparison implies that the force is potential, being stored up in the bent cross-bow until released by the finger on the trigger. None of the commentators seem to grasp the real point of the simile.

16. Amid the turmoil and tumult of battle, there may be seeming disorder and yet no real disorder at all; amid confusion and chaos, your array may be without head or tail, yet it will be proof against defeat.

形圓, literally "formation circular", is explained by Li Ch'üan as 無向背也 "without back or front." Mei Yao-ch'ên says: "The sub-divisions of the army having been previously fixed, and the various signals agreed upon, the separating and joining, the dispersing and collecting which will take place in the course of a battle, may give the appearance of disorder when no real disorder is possible. Your formation may be without head or tail, your dispositions all topsy-turvy, and yet a rout of your forces quite out of the question." It is a little difficult to decide whether 鬥亂 and 形圓 should not be taken as imperatives: "fight in disorder (for the purpose of deceiving the enemy), and you will be secure against real disorder." Cf. I. §20: 亂而取之.

17. 亂生於治怯生於勇弱生於彊
18. 治亂數也勇怯執也彊弱形也

17. Simulated disorder postulates perfect discipline; simulated fear postulates courage; simulated weakness postulates strength.

In order to make the translation intelligible, it is necessary to tone down the sharply paradoxical form of the original. Ts'ao Kung throws out a hint of the meaniug in his brief note: 皆毀形匿情也 "These things all serve to destroy formation and conceal one's condition." But Tu Mu is the first to put it quite plainly: "If you wish to feign confusion in order to lure the enemy on, you must first have perfect discipline; if you wish to display timidity in order to entrap the enemy, you must have extreme courage; if you wish to parade your weakness in order to make the enemy over-confident, you must have exceeding strength."

18. Hiding order beneath the cloak of disorder is simply a question of subdivision;

See *supra*, § 1.

concealing courage under a show of timidity presupposes a fund of latent energy;

It is passing strange that the commentators should understand 執 here as "circumstances" — a totally different sense from that which it has previously borne in this chapter. Thus Tu Mu says: 見有利之勢而不動敵人以我爲實怯也 "seeing that we are favourably circumstanced and yet make no move, the enemy will believe that we are really afraid."

masking strength with weakness is to be effected by tactical dispositions.

Chang Yü relates the following anecdote of Kao Tsu, the first Han Emperor: "Wishing to crush the Hsiung-nu, he sent out spies to report on their condition. But the Hsiung-nu, forewarned, carefully concealed all their able-bodied men and well-fed horses, and only allowed infirm soldiers and emaciated cattle to be seen. The result was that the spies one and all recommended the Emperor to deliver his attack. 婁敬 Lou Ching alone opposed them, saying: "When two countries go to war, they are naturally inclined to make an ostentatious display of their strength. Yet our spies have seen nothing but old age and infirmity. This is surely some *ruse* on the part of the enemy, and it would be unwise for us to attack." The Emperor, however, disregarding this advice, fell into the trap and found himself surrounded at 白登 Po-têng."

19. 故善動敵者形之敵必從之予之敵必取之
20. 以利動之以卒待之

19. Thus one who is skilful at keeping the enemy on the move maintains deceitful appearances, according to which the enemy will act.

Ts'ao Kung's note is 見羸形也 "Make a display of weakness and want," but Tu Mu rightly points out that 形 does not refer only to weakness: "If our force happens to be superior to the enemy's, weakness may be simulated in order to lure him on; but if inferior, he must be led to believe that we are strong, in order that he may keep off. In fact, all the enemy's movements should be determined by the signs that we choose to give him." The following anecdote of 孫臏 Sun Pin, a descendant of Sun Wu, is related at length in the 史記, chap. 65: In 341 B.C., the 齊 Ch'i State being at war with 魏 Wei, sent 田忌 T'ien Chi and Sun Pin against the general 龐涓 P'ang Chüan, who happened to be a deadly personal enemy of the latter. Sun Pin said: "The Ch'i State has a reputation for cowardice, and therefore our adversary despises us. Let us turn this circumstance to account." Accordingly, when the army had crossed the border into Wei territory, he gave orders to show 100,000 fires on the first night, 50,000 on the next, and the night after only 20,000. P'ang Chüan pursued them hotly, saying to himself: "I knew these men of Ch'i were cowards: their numbers have already fallen away by more than half." In his retreat, Sun Pin came to a narrow defile, which he calculated that his pursuers would reach after dark. Here he had a tree stripped of its bark, and inscribed upon it the words: "Under this tree shall P'ang Chüan die." Then, as night began to fall, he placed a strong body of archers in ambush near by, with orders to shoot directly they saw a light. Later on, P'ang Chüan arrived at the spot, and noticing the tree, struck a light in order to read what was written on it. His body was immediately riddled by a volley of arrows, and his whole army thrown into confusion. [The above is Tu Mu's version of the story; the *Shih Chi*, less dramatically but probably with more historical truth, makes P'ang Chüan cut his own throat with an exclamation of despair, after the rout of his army.]

He sacrifices something, that the enemy may snatch at it.

予 here = 與.

20. By holding out baits, he keeps him on the march; then with a body of picked men he lies in wait for him.

This would appear to be the meaning if we retain 卒, which Mei Yao-ch'ên explains as 精卒 "men of spirit." The *T'u Shu* reads 本,

21. 故善戰者求之於埶不責於人故能擇人而任埶

22. 任埶者其戰人也如轉木石木石之性安則靜危則動方則止圓則行

23. 故善戰人之埶如轉圓石於千仞之山者埶也

an emendation suggested by 李靖 Li Ching. The meaning then would be, "He lies in wait with the main body of his troops.".

21. The clever combatant looks to the effect of combined energy, and does not require too much from individuals.

Tu Mu says: "He first of all considers the power of his army in the bulk; afterwards he takes individual talent into account, and uses each man according to his capabilities. He does not demand perfection from the untalented."

Hence his ability to pick out the right men and to utilise combined energy.

Another reading has 之 instead of 埶. It would be interesting if Capt. Calthrop could tell us where the following occurs in the Chinese: "yet, when an opening or advantage shows, he pushes it to its limits."

22. When he utilises combined energy, his fighting men become as it were like unto rolling logs or stones. For it is the nature of a log or stone to remain motionless on level ground, and to move when on a slope; if four-cornered, to come to a standstill, but if round-shaped, to go rolling down.

Ts'ao Kung calls this 任自然埶 "the use of natural or inherent power." Capt. Calthrop ignores the last part of the sentence entirely. In its stead he has: "So await the opportunity, and so act when the opportunity arrives" — another absolutely gratuitous interpolation. The *T'ung Tien* omits 任.

23. Thus the energy developed by good fighting men is as the momentum of a round stone rolled down a mountain thousands of feet in height. So much on the subject of energy.

The *T'ung Tien* omits 善. The chief lesson of this chapter, in Tu Mu's opinion, is the paramount importance in war of rapid evolutions and sudden rushes. "Great results," he adds, "can thus be achieved with small forces."

VI. 虛實篇

1. 孫子曰凡先處戰地而待敵者佚後處戰地
而趨戰者勞
2. 故善戰者致人而不致於人

VI. WEAK POINTS AND STRONG.

Chang Yü attempts to explain the sequence of chapters as follows: "Chapter IV, on Tactical Dispositions, treated of the offensive and the defensive; chapter V, on Energy, dealt with direct and indirect methods. The good general acquaints himself first with the theory of attack and defence, and then turns his attention to direct and indirect methods. He studies the art of varying and combining these two methods before proceeding to the subject of weak and strong points. For the use of direct or indirect methods arises out of attack and defence, and the perception of weak and strong points depends again on the above methods. Hence the present chapter comes immediately after the chapter on Energy."

1. Sun Tzŭ said: Whoever is first in the field and awaits the coming of the enemy, will be fresh for the fight; whoever is second in the field and has to hasten to battle, will arrive exhausted.

Instead of 處, the *Yü Lan* has in both clauses the stronger word 據. For the antithesis between 佚 and 勞, cf. I. §23, where however 勞 is used as a verb.

2. Therefore the clever combatant imposes his will on the enemy, but does not allow the enemy's will to be imposed on him.

The next paragraph makes it clear that 致 does not merely mean, as Tu Mu says, 令敵來就我 "to make the enemy approach me," but rather to make him go in any direction I please. It is thus practically synonymous with 制. Cf. Tu Mu's own note on V. § 19. One mark of a great soldier is that he fights on his own terms or fights not at all. *

* See Col, Henderson's biography of Stonewall Jackson, 1902 ed., vol. II, p. 490.

3. 能使敵人自至者利之也能使敵人不得至
者害之也

4. 故敵佚能勞之飽能飢之安能動之

5. 出其所必趨趨其所不意

6. 行千里而不勞者行於無人之地也

3. By holding out advantages to him, he can cause the enemy to approach of his own accord; or, by inflicting damage, he can make it impossible for the enemy to draw near.

In the first case, he will entice him with a bait; in the second, he will strike at some important point which the enemy will have to defend.

4. If the enemy is taking his ease, he can harass him;

This passage may be cited as evidence against Mei Yao-Ch'ên's interpretation of I. § 23.

if well supplied with food, he can starve him out;

飢 is probably an older form than 饑, the reading of the original text. Both are given in the 説文.

if quietly encamped, he can force him to move.

The subject to 能 is still 善戰者; but these clauses would read better as direct admonitions, and in the next sentence we find Sun Tzŭ dropping insensibly into the imperative.

5. Appear at points which the enemy must hasten to defend; march swiftly to places where you are not expected.

The original text, adopted by the *T'u Shu*, has 出其所不趨; it has been altered to suit the context and the commentaries of Ts'ao Kung and Ho Shih, who evidently read 必趨. The other reading would mean: "Appear at points to which the enemy cannot hasten;" but in this case there is something awkward in the use of 趨. Capt. Calthrop is wrong of course with "appearing where the enemy is not."

6. An army may march great distances without distress, if it marches through country where the enemy is not.

We must beware of understanding 無人之地 as "uninhabited country." Sun Tzŭ habitually uses 人 in the sense of 敵, e. g. *supra*, § 2.

7. 攻而必取者攻其所不守也守而必固者守
其所不攻也

8. 故善攻者敵不知其所守善守者敵不知其
所攻

Ts'ao Kung sums up very well: 出空擊虛避其所守擊其
不意 "Emerge from the void [*q. d.* like "a bolt from the blue"], strike
at vulnerable points, shun places that are defended, attack in unexpected
quarters." The difference of meaning between 空 and 虛 is worth noting.

7. You can be sure of succeding in your attacks if you
only attack places which are undefended.

所不守 is of course hyperbolical; Wang Hsi rightly explains it as
"weak points; that is to say, where the general is lacking in capacity, or
the soldiers in spirit; where the walls are not strong enough, or the precautions not strict enough; where relief comes too late, or provisions are
too scanty, or the defenders are variance amongst themselves."

You can ensure the safety of your defence if you only
hold positions that cannot be attacked.

I. e., where there are none of the weak points mentioned above. There
is rather a nice point involved in the interpretation of this latter clause.
Tu Mu, Ch'ên Hao, and Mei Yao-ch'ên assume the meaning to be: "In
order to make your defence quite safe, you must defend *even* those places
that are not likely to be attacked;" and Tu Mu adds: "How much more,
then, those that will be attacked." Taken thus, however, the clause balances
less well with the preceding — always a consideration in the highly anti-
thetical style which is natural to the Chinese. Chang Yü, therefore, seems
to come nearer the mark in saying: "He who is skilled in attack flashes
forth from the topmost heights of heaven [see IV. § 7], making it impossible for the enemy to guard against him. This being so, the places that
I shall attack are precisely those that the enemy cannot defend ... He
who is skilled in defence hides in the most secret recesses of the earth,
making it impossible for the enemy to estimate his whereabouts. This
being so, the places that I shall hold are precisely those that the enemy
cannot attack."

8. Hence that general is skilful in attack whose opponent
does not know what to defend; and he is skilful in defence
whose opponent does not know what to attack.

An aphorism which puts the whole art of war into a nutshell.

9. 微乎微乎至於無形神乎神乎至於無聲故
能爲敵之司命

10. 進而不可禦者衝其虛也退而不可追者速
而不可及也

11. 故我欲戰敵雖高壘深溝不得不與我戰者
攻其所必救也

9. O divine art of sublety and secrecy! Through you we learn to be invisible, through you inaudible;

Literally, "without form or sound," but it is said of course with reference to the enemy. Chang Yü, whom I follow, draws no sharp distinction between 微 and 神, but Tu Mu and others think that 微 indicates the secrecy to be observed on the defensive, and 神 the rapidity to be displayed in attack. The *Yü Lan* text differs considerably from ours, reading: 微乎微乎故能隱於常形神乎神乎故能爲敵司命.

and hence we can hold the enemy's fate in our hands.

The *T'ung Tien* has 故能爲變化司命. Capt. Calthrop's version of this paragraph is so remarkable that I cannot refrain from quoting it in full: "Now the secrets of the art of offence are not to be easily apprehended, as a certain shape or noise can be understood, of the senses; but when these secrets are once learnt, the enemy is mastered."

10. You may advance and be absolutely irresistible, if you make for the enemy's weak points; you may retire and be safe from pursuit if your movements are more rapid than those of the enemy.

The second member of the sentence is weak, because 不可及 is nearly tautologous with 不可追. The *Yü Lan* reads 遠 for 速.

11. If we wish to fight, the enemy can be forced to an engagement even though he be sheltered behind a high rampart and a deep ditch. All we need do is to attack some other place that he will be obliged to relieve.

Tu Mu says: "If the enemy is the invading party, we can cut his line of communications and occupy the roads by which he will have to return; if we are the invaders, we may direct our attack against the sovereign himself." It is clear that Sun Tzŭ, unlike certain generals in the late Boer war, was no believer in frontal attacks.

12. 我不欲戰畫地而守之敵不得與我戰者乖
其所之也

13. 故形人而我無形則我專而敵分

12. If we do not wish to fight, we can prevent the enemy from engaging us even though the lines of our encampment be merely traced out on the ground. All we need do is to throw something odd and unaccountable in his way.

In order to preserve the parallelism with § 11, I should prefer to follow the *T'u Shu* text, which inserts 雖 before 畫地. This extremely concise expression is intelligibly paraphrased by Chia Lin: 雖未修壘塹 "even though we have constructed neither wall nor ditch." The real crux of the passage lies in 乖其所之也. 之 of course = 至. Ts'ao Kung defines 乖 by the word 戾, which is perhaps a case of *obscurum per obscurius*. Li Ch'üan, however, says: 設奇異而疑之 "we puzzle him by strange and unusual dispositions;" and Tu Mu finally clinches the meaning by three illustrative anecdotes — one of 諸葛亮 Chu-ko Liang, who when occupying 陽平 Yang-p'ing and about to be attacked by 司馬懿 Ssŭ-ma I, suddenly struck his colours, stopped the beating of the drums, and flung open the city gates, showing only a few men engaged in sweeping and sprinkling the ground. This unexpected proceeding had the intended effect; for Ssŭ-ma I, suspecting an ambush, actually drew off his army and retreated. What Sun Tzŭ is advocating here, therefore, is nothing more nor less than the timely use of "bluff." Capt. Calthrop translates: "and prevent the enemy from attacking by keeping him in suspense," which shows that he has not fully grasped the meaning of 乖.

13. By discovering the enemy's dispositions and remaining invisible ourselves, we can keep our forces concentrated, while the enemy's must be divided.

The conclusion is perhaps not very obvious, but Chang Yü (after Mei Yao-ch'ên) rightly explains it thus: "If the enemy's dispositions are visible, we can make for him in one body; whereas, our own dispositions being kept secret, the enemy will be obliged to divide his forces in order to guard against attack from every quarter." 形 is here used as an active verb: "to make to appear." See IV, note on heading. Capt. Calthrop's "making feints" is quite wrong.

14. 我專爲一敵分爲十是以十共其一也則我
衆而敵寡

15. 能以衆擊寡者則吾之所與戰者約矣

16. 吾所與戰之地不可知不可知則敵所備者
多敵所備者多則吾所與戰者寡矣

14. We can form a single united body, while the enemy must split up into fractions. Hence there will be a whole pitted against separate parts of a whole,

The original text has 以敵攻其一也, which in accordance with the *T'ung Tien* and *Yü Lan* has been altered as above. I adopt the more plausible reading of the *T'u Shu*: 是以十攻其一也, in spite of having to refer 十 to ourselves and not to the enemy. Thus Tu Yu and Mei Yao-ch'ên both regard 十 as the undivided force, consisting of so many parts, and 一 as each of the isolated fractions of the enemy. The alteration of 攻 into 共 can hardly be right, though the true text might conceivably have been 是以十共攻其一也.

which means that we shall be many to the enemy's few.

15. And if we are able thus to attack an inferior force with a superior one, our opponents will be in dire straits.

For 擊, the *T'ung Tien* and *Yü Lan* have 敵. Tu Yu, followed by the other commentators, arbitrarily defines 約 as 少而易勝 "few and easy to conquer," but only succeeds thereby in making the sentence absolutely pointless. As for Capt. Calthrop's translation: "In superiority of numbers there is economy of strength," its meaning is probably known to himself alone. In justification of my own rendering of 約, I would refer to *Lun Yü* IV. 2 and VII. 25 (3).

16. The spot where we intend to fight must not be made known; for then the enemy will have to prepare against a possible attack at several different points;

Sheridan once explained the reason of General Grant's victories by saying that "while his opponents were kept fully employed wondering what he was going to do, *he* was thinking most of what he was going to do himself."

and his forces being thus distributed in many directions, the numbers we shall have to face at any given point will be proportionately few.

17. 故備前則後寡備後則前寡備左則右寡備
右則左寡無所不備則無所不寡
18. 寡者備人者也眾者使人備己者也
19. 故知戰之地知戰之日則可千里而會戰

17. For should the enemy strengthen his van, he will weaken his rear; should he strengthen his rear, he will weaken his van; should he strengthen his left, he will weaken his right; should he strengthen his right, he will weaken his left. If he sends reinforcements everywhere, he will everywhere be weak.

In Frederick the Great's *Instructions to his Generals* we read: "A defensive war is apt to betray us into too frequent detachment. Those generals who have had but little experience attempt to protect every point, while those who are better acquainted with their profession, having only the capital object in view, guard against a decisive blow, and acquiesce in smaller misfortunes to avoid greater."

18. Numerical weakness comes from having to prepare against possible attacks; numerical strength, from compelling our adversary to make these preparations against us.

The highest generalship, in Col. Henderson's words, is "to compel the enemy to disperse his army, and then to concentrate superior force against each fraction in turn."

19. Knowing the place and the time of the coming battle, we may concentrate from the greatest distances in order to fight.

There is nothing about "defeating" anybody in this sentence, as Capt. Calthrop translates. What Sun Tzŭ evidently has in mind is that nice calculation of distances and that masterly employment of strategy which enable a general to divide his army for the purpose of a long and rapid march, and afterwards to effect a junction at precisely the right spot and the right hour in order to confront the enemy in overwhelming strength. Among many such successful junctions which military history records, one of the most dramatic and decisive was the appearance of Blücher just at the critical moment on the field of Waterloo.

20. 不知戰地不知戰日則左不能救右右不能
救左前不能救後後不能救前而況遠者數
十里近者數里乎

21. 以吾度之越人之兵雖多亦奚益於勝敗哉
故曰勝可為也

20. But if neither time nor place be known, then the left wing will be impotent to succour the right, the right equally impotent to succour the left, the van unable to relieve the rear, or the rear to support the van. How much more so if the furthest portions of the army are anything under a hundred *li* apart, and even the nearest are separated by several *li*!

The Chinese of this last sentence is a little lacking in precision, but the mental picture we are required to draw is probably that of an army advancing towards a given rendez-vous in separate columns, each of which has orders to be there on a fixed date. If the general allows the various detachments to proceed at haphazard, without precise instructions as to the time and place of meeting, the enemy will be able to annihilate the army in detail. Chang Yü's note may be worth quoting here: "If we do not know the place where our opponents mean to concentrate or the day on which they will join battle, our unity will be forfeited through our preparations for defence, and the positions we hold will be insecure. Suddenly happening upon a powerful foe, we shall be brought to battle in a flurried condition, and no mutual support will be possible between wings, vanguard or rear, especially if there is any great distance between the foremost and hindmost divisions of the army."

21. Though according to my estimate the soldiers of Yüeh exceed our own in number, that shall advantage them nothing in the matter of victory.

Capt. Calthrop omits 以吾度之, and his translation of the remainder is flabby and inaccurate. As Sun Tzŭ was in the service of the 吳 Wu State, it has been proposed to read 吳 instead of 吾 — a wholly unnecessary tampering with the text. Yüeh coincided roughly with the present province of Chehkiang. Li Ch'üan very strangely takes 越 not as the proper name, but in the sense of 過 "to surpass." No other commentator follows him. 勝敗 belongs to the class of expressions like 遠近 "distance," 大小 "magnitude," etc., to which the Chinese have to resort

4

22. 敵雖衆可使無鬥故策之而知得失之計
23. 作之而知動靜之理形之而知死生之地

in order to express abstract ideas of degree. The *T'u Shu*, however, omits 敗.

I say then that victory can be achieved.

Alas for these brave words! The long feud between the two states ended in 473 B.C. with the total defeat of Wu by 勾踐 Kou Chien and its incorporation in Yüeh. This was doubtless long after Sun Tzŭ's death. With his present assertion compare IV. § 4: 勝可知而不可爲 (which is the obviously mistaken reading of the *Yü Lan* here). Chang Yü is the only one to point out the seeming discrepancy, which he thus goes on to explain: "In the chapter on Tactical Dispositions it is said, 'One may *know* how to conquer without being able to *do* it,' whereas here we have the statement that 'victory can be achieved.' The explanation is, that in the former chapter, where the offensive and defensive are under discussion, it is said that if the enemy is fully prepared, one cannot make certain of beating him. But the present passage refers particularly to the soldiers of Yüeh who, according to Sun Tzŭ's calculations, will be kept in ignorance of the time and place of the impending struggle. That is why he says here that victory can be achieved."

22. Though the enemy be stronger in numbers, we may prevent him from fighting.

Capt. Calthrop quite unwarrantably translates: "*If* the enemy be many in number, prevent him," etc.

Scheme so as to discover his plans and the likelihood of their success.

This is the first of four similarly constructed sentences, all of which present decided difficulties. Chang Yü explains 知得失之計 as 知其計之得失. This is perhaps the best way of taking the words, though Chia Lin, referring 計 to ourselves and not the enemy, offers the alternative of 我得彼失之計皆先知也 "Know beforehand all plans conducive to our success and to the enemy's failure."

23. Rouse him, and learn the principle of his activity or inactivity.

Instead of 作, the *T'ung Tien*, *Yü Lan*, and also Li Ch'üan's text have 候, which the latter explains as "the observation of omens," and Chia Lin simply as "watching and waiting." 作 is defined by Tu Mu

24. 角之而知有餘不足之處
25. 故形兵之極至於無形無形則深閒不能窺
　　知者不能謀

as 激作, and Chang Yü tells us that by noting the joy or anger shown by the enemy on being thus disturbed, we shall be able to conclude whether his policy is to lie low or the reverse. He instances the action of Chu-ko Liang, who sent the scornful present of a woman's head-dress to Ssŭ-ma I, in order to goad him out of his Fabian tactics.

Force him to reveal himself, so as to find out his vulnerable spots.

Two commentators, Li Ch'üan and Chang Yü, take 形之 in the sense of 示之 "put on specious appearances." The former says: "You may either deceive the enemy by a show of weakness — striking your colours and silencing your drums; or by a show of strength — making a hollow display of camp-fires and regimental banners." And the latter quotes V. 19, where 形之 certainly seems to bear this sense. On the other hand, I would point to § 13 of this chapter, where 形 must with equal certainty be active. It is hard to choose between the two interpretations, but the context here agrees better, I think, with the one that I have adopted. Another difficulty arises over 死生之地, which most of the commentators, thinking no doubt of the 死地 in XI. § 1, refer to the actual *ground* on which the enemy is encamped. The notes of Chia Lin and Mei Yao-ch'ên, however, seem to favour my view. The same phrase has a somewhat different meaning in I. § 2.

24. **Carefully compare the opposing army with your own,**

Tu Yu is right, I think, in attributing this force to 角; Ts'ao Kung defines it simply as 量. Capt. Calthrop surpasses himself with the staggering translation "Flap the wings"! Can the Latin *cornu* (in its figurative sense) have been at the back of his mind?

so that you may know where strength is superabundant and where it is deficient.

Cf. IV. § 6.

25. **In making tactical dispositions, the highest pitch you can attain is to conceal them;**

The piquancy of the paradox evaporates in translation. 無形 is perhaps not so much actual invisibility (see *supra*, § 9) as "showing no sign" of what you mean to do, of the plans that are formed in your brain.

26. 因形而錯勝於眾眾不能知

27. 人皆知我所以勝之形而莫知吾所以制勝
之形

28. 故其戰勝不復而應形於無窮

conceal your dispositions, and you will be safe from the
prying of the subtlest spies, from the machinations of the
wisest brains.

深閒 is expanded by Tu Mu into 雖有閒者深來窺我.
[For 閒, see XIII, note on heading.] He explains 知者 in like fashion:
雖有智能之士亦不能謀我也 "though the enemy may
have clever and capable officers, they will not be able to lay any plans
against us."

26. How victory may be produced for them out of the
enemy's own tactics — that is what the multitude cannot
comprehend.

All the commentators except Li Ch'üan make 形 refer to the enemy.
So Ts'ao Kung: 因敵形而立勝. 錯 is defined as 置. The
Tu Shu has 措, with the same meaning. See IV. § 13. The *Yü Lan*
reads 作, evidently a gloss.

27. All men can see the tactics whereby I conquer,
but what none can see is the strategy out of which victory
is evolved.

I. e., everybody can see superficially how a battle is won; what they
cannot see is the long series of plans and combinations which has preceded
the battle. It seems justifiable, then, to render the first 形 by "tactics"
and the second by "strategy."

28. Do not repeat the tactics which have gained you
one victory, but let your methods be regulated by the
infinite variety of circumstances.

As Wang Hsi sagely remarks: "There is but one root-principle (理)
underlying victory, but the tactics (形) which lead up to it are infinite
in number." With this compare Col. Henderson: "The rules of strategy
are few and simple. They may be learned in a week. They may be
taught by familiar illustrations or a dozen diagrams. But such knowledge
will no more teach a man to lead an army like Napoleon than a knowledge
of grammar will teach him to write like Gibbon."

29. 夫兵形象水水之行避高而趨下
30. 兵之形避實而擊虛
31. 水因地而制流兵因敵而制勝
32. 故兵無常勢水無常形
33. 能因敵變化而取勝者謂之神
34. 故五行無常勝四時無常位日有短長月有
死生

29. Military tactics are like unto water; for water in its natural course runs away from high places and hastens downwards.

行 is 劉晝子 Liu Chou-tzŭ's reading for 形 in the original text.

30. So in war, the way is to avoid what is strong and to strike at what is weak.

Like water, taking the line of least resistance.

31. Water shapes its course according to the nature of the ground over which it flows;

The *T'ung Tien* and *Yü Lan* read 制形, — the latter also 制行. The present text is derived from Cheng Yu-hsien.

the soldier works out his victory in relation to the foe whom he is facing.

32. Therefore, just as water retains no constant shape, so in warfare there are no constant conditions.

33. He who can modify his tactics in relation to his opponent and thereby succeed in winning, may be called a heaven-born captain.

34. The five elements

Water, fire, wood, metal, earth.

are not always equally predominant;

That is, as Wang Hsi says: 迭相克也 "they predominate alternately."

the four seasons make way for each other in turn.

Literally, "have no invariable seat."

There are short days and long; the moon has its periods of waning and waxing.

Cf. V. § 6. The purport of the passage is simply to illustrate the want of fixity in war by the changes constantly taking place in Nature. The comparison is not very happy, however, because the regularity of the phenomena which Sun Tzŭ mentions is by no means paralleled in war.

VII. 軍爭篇

1. 孫子曰凡用兵之法將受命於君
2. 合軍聚衆交和而舍

VII. MANŒUVRING.

The commentators, as well as the subsequent text, make it clear that this is the real meaning of 軍爭. Thus, Li Ch'üan says that 爭 means 趨利 "marching rapidly to seize an advantage"; Wang Hsi says: 爭者爭利得利則勝 "'Striving' means striving for an advantage; this being obtained, victory will follow;" and Chang Yü: 兩軍相對而爭利也 "The two armies face to face, and each striving to obtain a tactical advantage over the other." According to the latter commentator, then, the situation is analogous to that of two wrestlers manœuvring for a "hold," before coming to actual grips. In any case, we must beware of translating 爭 by the word "fighting" or "battle," as if it were equivalent to 戰. Capt. Calthrop falls into this mistake.

1. Sun Tzŭ said: In war, the general receives his commands from the sovereign.

For 君 there is another reading 天, which Li Ch'üan explains as 恭行天罰 "being the reverent instrument of Heaven's chastisement."

2. Having collected an army and concentrated his forces, he must blend and harmonise the different elements thereof before pitching his camp.

Ts'ao Kung takes 和 as referring to the 和門 or main gate of the military camp. This, Tu Mu tells us, was formed with a couple of flags hung across. [Cf. *Chou Li*, ch. xxvii. fol. 31 of the Imperial edition: 直旌門.] 交和 would then mean "setting up his 和門 opposite that of the enemy." But Chia Lin's explanation, which has been adopted

3. 莫難於軍爭軍爭之難者以迂爲直以患爲利

above, is on the whole simpler and better. Chang Yü, while following Ts'ao Kung, adds that the words may also be taken to mean "the establishment of harmony and confidence between the higher and lower ranks before venturing into the field;" and he quotes a saying of Wu Tzŭ (chap. I *ad init.*): "Without harmony in the State, no military expedition can be undertaken; without harmony in the army, no battle array can be formed." In the historical romance 東周列國, chap. 75, Sun Tzŭ himself is represented as saying to 伍員 Wu Yüan: 大凡行兵之法先除內患然後方可外征 "As a general rule, those who are waging war should get rid of all domestic troubles before proceeding to attack the external foe." 舍 is defined as 止. It here conveys the notion of encamping after having taken the field.

3. After that, comes tactical manœuvring, than which there is nothing more difficult.

I have departed slightly from the traditional interpretation of Ts'ao Kung, who says: 從始受命至於交和軍爭難也 "From the time of receiving the sovereign's instructions until our encampment over against the enemy, the tactics to be pursued are most difficult." It seems to me that the 軍爭 tactics or manœuvres can hardly be said to begin until the army has sallied forth and encamped, and Ch'ên Hao's note gives colour to this view: "For levying, concentrating, harmonising and intrenching an army, there are plenty of old rules which will serve. The real difficulty comes when we engage in tactical operations." Tu Yu also observes that "the great difficulty is to be beforehand with the enemy in seizing favourable positions."

The difficulty of tactical manœuvring consists in turning the devious into the direct, and misfortune into gain.

以迂爲直 is one of those highly condensed and somewhat enigmatical expressions of which Sun Tzŭ is so fond. This is how it is explained by Ts'ao Kung: 示以遠速其道里先敵至也 "Make it appear that you are a long way off, then cover the distance rapidly and arrive on the scene before your opponent." Tu Mu says: "Hoodwink the enemy, so that he may be remiss and leisurely while you are dashing along with the utmost speed." Ho Shih gives a slightly different turn to the sentence: "Although you may have difficult ground to traverse and natural obstacles to encounter, this is a drawback which

4. 故迂其途而誘之以利後人發先人至此知
迂直之計者也

can be turned into actual advantage by celerity of movement." Signal examples of this saying are afforded by the two famous passages across the Alps — that of Hannibal, which laid Italy at his mercy, and that of Napoleon two thousand years later, which resulted in the great victory of Marengo.

4. Thus, to take a long and circuitous route, after enticing the enemy out of the way, and though starting after him, to contrive to reach the goal before him, shows knowledge of the artifice of *deviation*.

Chia Lin understands 途 as the *enemy's* line of march, thus: "If our adversary's course is really a short one, and we can manage to divert him from it (迂之) either by simulating weakness or by holding out some small advantage, we shall be able to beat him in the race for good positions." This is quite a defensible view, though not adopted by any other commentator. 人 of course = 敵, and 後 and 先 are to be taken as verbs. Tu Mu cites the famous march of 趙奢 Chao Shê in 270 B.C. to relieve the town of 關與 O-yü, which was closely invested by a 秦 Ch'in army. [It should be noted that the above is the correct pronunciation of 關與, as given in the commentary on the *Ch'ien Han Shu*, ch. 34. Giles' dictionary gives "Yü-yü," and Chavannes, I know not on what authority, prefers to write "Yen-yü." The name is omitted altogether from Playfair's "Cities and Towns."] The King of Chao first consulted 廉頗 Lien P'o on the advisability of attempting a relief, but the latter thought the distance too great, and the intervening country too rugged and difficult. His Majesty then turned to Chao Shê, who fully admitted the hazardous nature of the march, but finally said: "We shall be like two rats fighting in a hole — and the pluckier one will win!" So he left the capital with his army, but had only gone a distance of 30 *li* when he stopped and began throwing up intrenchments. For 28 days he continued strengthening his fortifications, and took care that spies should carry the intelligence to the enemy. The Ch'in general was overjoyed, and attributed his adversary's tardiness to the fact that the beleaguered city was in the Han State, and thus not actually part of Chao territory. But the spies had no sooner departed than Chao Shê began a forced march lasting for two days and one night, and arrived on the scene of action with such astonishing rapidity that he was able to occupy a commanding position on the 北山 "North hill" before the enemy

5. 故軍爭爲利衆爭爲危

6. 舉軍而爭利則不及委軍而爭利則輜重捐

7. 是故卷甲而趨日夜不處倍道兼行百里而
爭利則擒三將軍

had got wind of his movements. A crushing defeat followed for the Ch'in forces, who were obliged to raise the siege of O-yü in all haste and retreat across the border. [See 史記, chap. 81.]

5. Manœuvring with an army is advantageous; with an undisciplined multitude, most dangerous.

I here adopt the reading of the *T'ung Tien*, Chêng Yu-hsien and the *T'u Shu*, where 衆 appears to supply the exact *nuance* required in order to make sense. The standard text, on the other hand, in which 軍 is repeated, seems somewhat pointless. The commentators take it to mean that manœuvres may be profitable, or they may be dangerous: it all depends on the ability of the general. Capt. Calthrop translates 衆爭 "the wrangles of a multitude"!

6. If you set a fully equipped army in march in order to snatch an advantage, the chances are that you will be too late.

The original text has 故 instead of 舉; but a verb is needed to balance 委.

On the other hand, to detach a flying column for the purpose involves the sacrifice of its baggage and stores.

委軍 is evidently unintelligible to the Chinese commentators, who paraphrase the sentence as though it began with 棄輜. Absolute tautology in the apodosis can then only be avoided by drawing an impossibly fine distinction between 棄 and 捐. I submit my own rendering without much enthusiasm, being convinced that there is some deep-seated corruption in the text. On the whole, it is clear that Sun Tzŭ does not approve of a lengthy march being undertaken without supplies. Cf. *infra*, § 11.

7. Thus, if you order your men to roll up their buff-coats,

卷甲 does not mean "to discard one's armour," as Capt. Calthrop translates, but implies on the contrary that it is to be carried with you. Chang Yü says: 猶悉甲也 "This means, in full panoply."

8. 勁者先罷者後其法十一而至

9. 五十里而爭利則蹶上將軍其法半至

10. 三十里而爭利則三分之二至

and make forced marches without halting day or night, covering double the usual distance at a stretch,

The ordinary day's march, according to Tu Mu, was 30 *li*; but on one occasion, when pursuing 劉備 Liu Pei, Ts'ao Ts'ao is said to have covered the incredible distance of 300 *li* within twenty-four hours.

doing a hundred *li* in order to wrest an advantage, the leaders of all your three divisions will fall into the hands of the enemy.

8. The stronger men will be in front, the jaded ones will fall behind, and on this plan only one-tenth of your army will reach its destination.

For 罷, see II. § 14. The moral is, as Ts'ao Kung and others point out: Don't march a hundred *li* to gain a tactical advantage, either with or without impedimenta. Manœuvres of this description should be confined to short distances. Stonewall Jackson said: "The hardships of forced marches are often more painful than the dangers of battle." He did not often call upon his troops for extraordinary exertions. It was only when he intended a surprise, or when a rapid retreat was imperative, that he sacrificed everything to speed.

9. If you march fifty *li* in order to outmanœuvre the enemy, you will lose the leader of your first division, and only half your force will reach the goal.

蹶 is explained as similar in meaning to 挫: literally, "the leader of the first division will be *torn away*." Cf. Tso Chuan, 襄 19th year: 是謂蹶其本 "This is a case of [the falling tree] tearing up its roots."

10. If you march thirty *li* with the same object, two-thirds of your army will arrive.

In the *T'ung Tien* is added: 以是知軍爭之難 "From this we may know the difficulty of manœuvring."

---

* See Col. Henderson, *op. cit.* vol. I. p. 426.

11. 是故軍無輜重則亡無糧食則亡無委積則亡

12. 故不知諸侯之謀者不能豫交

13. 不知山林險阻沮澤之形者不能行軍

14. 不用鄉導者不能得地利

11. We may take it then that an army without its baggage-train is lost; without provisions it is lost; without bases of supply it is lost.

委積 is explained by Tu Yu as 芻草之屬 "fodder and the like;" by Tu Mu and Chang Yü as 財貨 "goods in general;" and by Wang Hsi as 薪鹽蔬材之屬 "fuel, salt, foodstuffs, etc." But I think what Sun Tzŭ meant was "stores accumulated in dépôts," as distinguished from 輜重 and 糧食, the various impedimenta accompanying an army on its march. Cf. *Chou Li*, ch. xvi. fol. 10: 委人...斂薪芻凡疏材木材凡畜聚之物.

12. We cannot enter into alliances until we are acquainted with the designs of our neighbours.

豫＝先. Li Ch'üan understands it as 備 "guard against," which is hardly so good. An original interpretation of 交 is given by Tu Mu, who says it stands for 交兵 or 合戰 "join in battle."

13. We are not fit to lead an army on the march unless we are familiar with the face of the country — its mountains and forests, its pitfalls

險, defined as 坑塹 (Ts'ao Kung) or 坑坎 (Chang Yü).

and precipices,

阻, defined as 一高一下.

its marshes

沮, defined as 水草漸洳者.

and swamps.

澤, defined as 衆水所歸而不流者.

14. We shall be unable to turn natural advantages to account unless we make use of local guides.

§§ 12—14 are repeated in chap. XI. § 52.

15. 故兵以詐立以利動
16. 以分合爲變者也
17. 故其疾如風其徐如林
18. 侵掠如火不動如山

15. In war, practise dissimulation, and you will succeed.

According to Tu Mu, 立 stands for 立勝. Cf. I. § 18. In the tactics of Turenne, deception of the enemy, especially as to the numerical strength of his troops, took a very prominent position. *

Move only if there is a real advantage to be gained.

This is the interpretation of all the commentators except Wang Hsi, who has the brief note 誘之也 "Entice out the enemy" (by offering him some apparent advantage).

16. **Whether to concentrate or to divide your troops, must be decided by circumstances.**

17. Let your rapidity be that of the wind,

The simile is doubly appropriate, because the wind is not only swift but, as Mei Yao-ch'ên points out, 無形跡 "invisible and leaves no tracks."

your compactness that of the forest.

It is hardly possible to take 徐 here in its ordinary sense of "sedate," as Tu Yu tries to do. Mêng Shih comes nearer the mark in his note 緩行須有行列 "When slowly marching, order and ranks must be preserved" — so as to guard against surprise attacks. But natural forests do not grow in rows, whereas they do generally possess the quality of density or compactness. I think then that Mei Yao-ch'ên uses the right adjective in saying 如林之森然.

18. In raiding and plundering be like fire,

Cf. *Shih Ching*, IV. 3. iv. 6: 如火烈烈則莫我敢曷 "Fierce as a blazing fire which no man can check."

in immovability like a mountain.

That is, when holding a position from which the enemy is trying to dislodge you, or perhaps, as Tu Yu says, when he is trying to entice you into a trap.

---

* For a number of maxims on this head, see "Marshal Turenne" (Longmans, 1907), p. 29.

19. 難知如陰動如雷霆
20. 掠鄉分眾廓地分利

19. Let your plans be dark and impenetrable as night, and when you move, fall like a thunderbolt.

The original text has 震 instead of 霆. Cf. IV. § 7. Tu Yu quotes a saying of T'ai Kung which has passed into a proverb: 疾雷不及掩耳疾電不及瞑目 "You cannot shut your ears to the thunder or your eyes to the lightning — so rapid are they." Likewise, an attack should be made so quickly that it cannot be parried.

20. When you plunder a countryside, let the spoil be divided amongst your men;

The reading of Tu Yu, Chia Lin, and apparently Ts'ao Kung, is 指向分眾, which is explained as referring to the subdivision of the army, mentioned in V. §§ 1, 2, by means of banners and flags, serving to point out (指) to each man the way he should go (向). But this is very forced, and the ellipsis is too great, even for Sun Tzŭ. Luckily, the *T'ung Tien* and *Yü Lan* have the variant 嚮, which not only suggests the true reading 鄉, but affords some clue to the way in which the corruption arose. Some early commentator having inserted 向 as the sound of 鄉, the two may afterwards have been read as one character; and this being interchangeable with 向, 鄉 must finally have disappeared altogether. Meanwhile, 掠 would have been altered to 指 in order to make sense. As regards 分眾, I believe that Ho Shih alone has grasped the real meaning, the other commentators understanding it as "dividing the men into parties" to search for plunder. Sun Tzŭ wishes to lessen the abuses of indiscriminate plundering by insisting that all booty shall be thrown into a common stock, which may afterwards be fairly divided amongst all.

when you capture new territory, cut it up into allotments for the benefit of the soldiery.

That this is the meaning, may be gathered from Tu Mu's note: 開土拓境則分割與有功者. The 三略 gives the same advice: 獲地裂之: 廓 means "to enlarge" or "extend" — at the expense of the enemy, understood. Cf. *Shih Ching*, III. 1. vii. 1: 憎其式廓 "hating all the great States." Ch'ên Hao also says 屯兵種蒔 "quarter your soldiers on the land, and let them sow and plant

21. 懸權而動
22. 先知迂直之計者勝此軍爭之法也
23. 軍政曰言不相聞故爲金鼓視不相見故爲
旌旗

it." It is by acting on this principle, and harvesting the lands they invaded, that the Chinese have succeeded in carrying out some of their most memorable and triumphant expeditions, such as that of 班超 Pan Ch'ao who penetrated to the Caspian, and in more recent years, those of 福康安 Fu-k'ang-an and 左宗棠 Tso Tsung-t'ang.

21. Ponder and deliberate

Note that both these words, like the Chinese 懸權, are really metaphors derived from the use of scales.

before you make a move.

Chang Yü quotes 尉繚子 as saying that we must not break camp until we have gauged the resisting power of the enemy and the cleverness of the opposing general. Cf. the "seven comparisons" in I. § 13. Capt. Calthrop omits this sentence.

22. He will conquer who has learnt the artifice of deviation.

See *supra*, §§ 3, 4.

Such is the art of manœuvring.

With these words, the chapter would naturally come to an end. But there now follows a long appendix in the shape of an extract from an earlier book on War, now lost, but apparently extant at the time when Sun Tzŭ wrote. The style of this fragment is not noticeably different from that of Sun Tzŭ himself, but no commentator raises a doubt as to its genuineness.

23. The Book of Army Management says:

It is perhaps significant that none of the earlier commentators give us any information about this work. Mei Yao-Ch'ên calls it 軍之舊典 "an ancient military classic," and Wang Hsi, 古軍書 "an old book on war." Considering the enormous amount of fighting that had gone on for centuries before Sun Tzŭ's time between the various kingdoms and principalities of China, it is not in itself improbable that a collection of military maxims should have been made and written down at some earlier period.

24. 夫金鼓旌旗者所以一民之耳目也

25. 民既專一則勇者不得獨進怯者不得獨退
此用衆之法也

On the field of battle,

Implied, though not actually in the Chinese.

the spoken word does not carry far enough: hence the institution of gongs and drums.

I have retained the words 金鼓 of the original text, which recur in the next paragraph, in preference to the other reading 鼓鐸 "drums and bells," which is found in the *T'ung Tien*, *Pei T'ang Shu Ch'ao* and *Yü Lan*. 鐸 is a bell with a clapper. See *Lun Yü* III. 24, *Chou Li* XXIX. 15, 29. 金 of course would include both gongs and bells of every kind. The *T'u Shu* inserts a 之 after each 爲.

Nor can ordinary objects be seen clearly enough: hence the institution of banners and flags.

24. Gongs and drums, banners and flags, are means whereby the ears and eyes of the host

The original text, followed by the *T'u Shu*, has 人 for 民 here and in the next two paragraphs. But, as we have seen, 人 is generally used in Sun Tzǔ for the enemy.

may be focussed on one particular point.

Note the use of 一 as a verb. Chang Yü says: 視聽均齊則雖百萬之衆進退如一矣 "If sight and hearing converge simultaneously on the same object, the evolutions of as many as a million soldiers will be like those of a single man"!

25. The host thus forming a single united body, it is impossible either for the brave to advance alone, or for the cowardly to retreat alone.

Chang Yü quotes a saying: 令不進而進與令不退而退厥罪惟均 "Equally guilty are those who advance against orders and those who retreat against orders." Tu Mu tells a story in this connection of 吳起 Wu Ch'i, when he was fighting against the Ch'in State. Before the battle had begun, one of his soldiers, a man of matchless daring, sallied forth by himself, captured two heads from the enemy, and

26. 故夜戰多火鼓畫戰多旌旗所以變民之耳
目也

27. 故三軍可奪氣將軍可奪心

returned to camp. Wu Ch'i had the man instantly executed, whereupon an officer ventured to remonstrate, saying: "This man was a good soldier, and ought not to have been beheaded." Wu Ch'i replied: "I fully believe he was a good soldier, but I had him beheaded because he acted without orders."

This is the art of handling large masses of men.

26. In night-fighting, then, make much use of signal-fires and drums, and in fighting by day, of flags and banners, as a means of influencing the ears and eyes of your army.

The *T'ung Tien* has the bad variant 便 for 變. With regard to the latter word, I believe I have hit off the right meaning, the whole phrase being slightly elliptical for "influencing the movements of the army through their senses of sight and hearing." Li Ch'üan, Tu Mu and Chia Lin certainly seem to understand it thus. The other commentators, however, take 民 (or 人) as the enemy, and 變 as equivalent to 變惑 or 變亂 "to perplex" or "confound." This does not agree so well with what has gone before, though on the other hand it renders the transition to § 27 less abrupt. The whole question, I think, hinges on the alternative readings 民 and 人. The latter would almost certainly denote the enemy. Ch'ên Hao alludes to 李光弼 Li Kuang-pi's night ride to 河陽 Ho-yang at the head of 500 mounted men; they made such an imposing display with torches, that though the rebel leader 史思明 Shih Ssŭ-ming had a large army, he did not dare to dispute their passage. [Ch'ên Hao gives the date as 天寶末 A.D. 756; but according to the 新唐書 New T'ang History, 列傳 61, it must have been later than this, probably 760.]

27. A whole army may be robbed of its spirit;

"In war," says Chang Yü, "if a spirit of anger can be made to pervade all ranks of an army at one and the same time, its onset will be irresistible. Now the spirit of the enemy's soldiers will be keenest when they have newly arrived on the scene, and it is therefore our cue not to fight at once, but to wait until their ardour and enthusiasm have worn off, and then strike. It is in this way that they may be robbed of their keen spirit." Li Ch'üan and others tell an anecdote (to be found in the *Tso Chuan,*

5

28. 是故朝氣銳晝氣惰暮氣歸

29. 故善用兵者避其銳氣擊其惰歸此治氣者
也

莊公 year 10, § 1) of 曹劌 Ts'ao Kuei, a *protégé* of Duke Chuang of Lu. The latter State was attacked by Ch'i, and the Duke was about to join battle at 長勺 Ch'ang-cho, after the first roll of the enemy's drums, when Ts'ao said: "Not just yet." Only after their drums had beaten for the third time, did he give the word for attack. Then they fought, and the men of Ch'i were utterly defeated. Questioned afterwards by the Duke as to the meaning of his delay, Ts'ao Kuei replied: "In battle, a courageous spirit is everything. Now the first roll of the drum tends to create this spirit, but with the second it is already on the wane, and after the third it is gone altogether. I attacked when their spirit was gone and ours was at its height. Hence our victory." 吳子 (chap. 4) puts "spirit" first among the "four important influences" in war, and continues: 三軍之衆百萬之師張設輕重在於一人是謂氣機 "The value of a whole army — a mighty host of a million men — is dependent on one man alone: such is the influence of spirit!"

a commander-in-chief may be robbed of his presence of mind.

Capt. Calthrop goes woefully astray with "defeat his general's ambition." Chang Yü says: 心者將之所主也夫治亂勇怯皆主於心 "Presence of mind is the general's most important asset. It is the quality which enables him to discipline disorder and to inspire courage into the panic-stricken." The great general 李靖 Li Ching (A.D. 571—649) has a saying: 夫攻者不止攻其城擊其陳而已必有攻其心之術焉 "Attacking does not merely consist in assaulting walled cities or striking at an army in battle array; it must include the art of assailing the enemy's mental equilibrium." [問對, pt. 3.]

28. Now a soldier's spirit is keenest in the morning;

Always provided, I suppose, that he has had breakfast. At the battle of the Trebia, the Romans were foolishly allowed to fight fasting, whereas Hannibal's men had breakfasted at their leisure. See Livy, XXI, liv. 8, lv. 1 and 8.

by noonday it has begun to flag; and in the evening, his mind is bent only on returning to camp.

29. A clever general, therefore,

30. 以治待亂以靜待譁此治心者也
31. 以近待遠以佚待勞以飽待飢此治力者也
32. 無要正正之旗勿擊堂堂之陳此治變者也

The 故, which certainly seems to be wanted here, is omitted in the *T'u Shu*.

avoids an army when its spirit is keen, but attacks it when it is sluggish and inclined to return. This is the art of studying moods.

The *T'ung Tien*, for reasons of 避諱 "avoidance of personal names of the reigning dynasty," reads 理 for 治 in this and the two next paragraphs.

30. Disciplined and calm, to await the appearance of disorder and hubbub amongst the enemy: — this is the art of retaining self-possession.

31. To be near the goal while the enemy is still far from it, to wait at ease

The *T'ung Tien* has 逸 for 佚. The two characters are practically synonymous, but according to the commentary, the latter is the form always used in Sun Tzŭ.

while the enemy is toiling and struggling, to be well-fed while the enemy is famished: — this is the art of husbanding one's strength.

32. To refrain from intercepting

邀 is the reading of the original text. But the 兵書要訣 quotes the passage with 要 *yao*[1] (also meaning "to intercept"), and this is supported by the *Pei T'ang Shu Ch'ao*, the *Yü Lan*, and Wang Hsi's text.

an enemy whose banners are in perfect order, to refrain from attacking an army drawn up in calm and confident array:

For this translation of 堂堂, I can appeal to the authority of Tu Mu, who defines the phrase as 無懼. The other commentators mostly follow Ts'ao Kung, who says 大, probably meaning "grand and imposing". Li Ch'üan, however, has 部分 "in subdivisions," which is somewhat strange.

33. 故用兵之法高陵勿向背邱勿逆
34. 佯北勿從銳卒勿攻
35. 餌兵勿食歸師勿遏

— this is the art of studying circumstances.

I have not attempted a uniform rendering of the four phrases 治氣,
治心, 治力 and 治變, though 治 really bears the same meaning
in each case. It is to be taken, I think, not in the sense of "to govern"
or "control," but rather, as K'ang Hsi defines it, = 簡習 "to examine
and practise," hence "look after," "keep a watchful eye upon." We may
find an example of this use in the *Chou Li*, XVIII. fol. 46: 治其大禮.
Sun Tzŭ has not told us to control or restrain the quality which he calls
氣, but only to observe the time at which it is strongest. As for 心,
it is important to remember that in the present context it can only mean
"presence of mind." To speak of "controlling presence of mind" is absurd, and Capt. Calthrop's "to have the heart under control" is hardly
less so. The whole process recommended here is that of VI. § 2:
致人而不致於人.

33. It is a military axiom not to advance uphill against
the enemy, nor to oppose him when he comes downhill.

The *Yü Lan* reads 倍 for 背.

34. Do not pursue an enemy who simulates flight; do
not attack soldiers whose temper is keen.

35. Do not swallow a bait offered by the enemy.

Li Ch'üan and Tu Mu, with extraordinary inability to see a metaphor,
take these words quite literally of food and drink that have been poisoned
by the enemy. Ch'ên Hao and Chang Yü carefully point out that the
saying has a wider application. The *T'ung Tien* reads 貪 "to covet"
instead of 食. The similarity of the two characters sufficiently accounts
for the mistake.

Do not interfere with an army that is returning home.

The commentators explain this rather singular piece of advice by saying
that a man whose heart is set on returning home will fight to the death
against any attempt to bar his way, and is therefore too dangerous an
opponent to be tackled. Chang Yü quotes the words of Han Hsin:
從思東歸之士何所不克 "Invincible is the soldier who
hath his desire and returneth homewards." A marvellous tale is told of

36. 圍師必闕窮寇勿迫

Ts'ao Ts'ao's courage and resource in ch. 1 of the *San Kuo Chih*, 武帝紀: In 198 A.D., he was besieging 張繡 Chang Hsiu in 穰 Jang, when 劉表 Liu Piao sent reinforcements with a view to cutting off Ts'ao's retreat. The latter was obliged to draw off his troops, only to find himself hemmed in between two enemies, who were guarding each outlet of a narrow pass in which he had engaged himself. In this desperate plight Ts'ao waited until nightfall, when he bored a tunnel into the mountain side and laid an ambush in it. Then he marched on with his baggage-train, and when it grew light, Chang Hsiu, finding that the bird had flown, pressed after him in hot pursuit. As soon as the whole army had passed by, the hidden troops fell on its rear, while Ts'ao himself turned and met his pursuers in front, so that they were thrown into confusion and annihilated. Ts'ao Ts'ao said afterwards: 虜遏吾歸師而與吾死地戰吾是以知勝矣 "The brigands tried to check my army in its retreat and brought me to battle in a desperate position: hence I knew how to overcome them."

36. When you surround an army, leave an outlet free.

This does not mean that the enemy is to be allowed to escape. The object, as Tu Mu puts it, is 示以生路令無必死之心 "to make him believe that there is a road to safety, and thus prevent his fighting with the courage of despair." Tu Mu adds pleasantly: 因而擊之 "After that, you may crush him."

Do not press a desperate foe too hard.

For 迫, the *T'u Shu* reads 追 "pursue." Ch'ên Hao quotes the saying: 鳥窮則博獸窮則噬 "Birds and beasts when brought to bay will use their claws and teeth." Chang Yü says: 敵若焚舟破釜決一戰則不可逼迫來 "If your adversary has burned his boats and destroyed his cooking-pots, and is ready to stake all on the issue of a battle, he must not be pushed to extremities." The phrase 窮寇 doubtless originated with Sun Tzŭ. The *P'ei Wên Yün Fu* gives four examples of its use, the earliest being from the *Ch'ien Han Shu*, and I have found another in chap. 34 of the same work. Ho Shih illustrates the meaning by a story taken from the life of 符彥卿 Fu Yen-ch'ing in ch. 251 of the 宋史. That general, together with his colleague 杜重威 Tu Chung-wei, was surrounded by a vastly superior army of Khitans in the year 945 A.D. The country was bare and desert-like, and the little Chinese force was soon in dire straits for want of

37. 此用兵之法也

water. The wells they bored ran dry, and the men were reduced to squeezing lumps of mud and sucking out the moisture. Their ranks thinned rapidly, until at last Fu Yen-ch'ing exclained: "We are desperate men. Far better to die for our country than to go with fettered hands into captivity!" A strong gale happened to be blowing from the northeast and darkening the air with dense clouds of sandy dust. Tu Chungwei was for waiting until this had abated before deciding on a final attack; but luckily another officer, 李守貞 Li Shou-chêng by name, was quicker to see an opportunity, and said: "They are many and we are few, but in the midst of this sandstorm our numbers will not be discernible; victory will go to the strenuous fighter, and the wind will be our best ally." Accordingly, Fu Yen-ch'ing made a sudden and wholly unexpected onslaught with his cavalry, routed the barbarians and succeeded in breaking through to safety. [Certain details in the above account have been added from the 歷代紀事年表, ch. 78.]

37. Such is the art of warfare.

Chêng Yu-hsien in his 遺說 inserts 妙 after 法. I take it that these words conclude the extract from the 軍政 which began at § 23.

VIII. 九變篇

1. 孫子曰凡用兵之法將受命於君合軍聚衆
2. 圮地無舍衢地合交絕地無留圍地則謀死地則戰

VIII. VARIATION OF TACTICS.

The heading means literally "The Nine Variations," but as Sun Tzŭ does not appear to enumerate these, and as, indeed, he has already told us (V. §§ 6—11) that such deflections from the ordinary course are practically innumerable, we have little option but to follow Wang Hsi, who says that "Nine" stands for an indefinitely large number. "All it means is that in warfare 當極其變 we ought to vary our tactics to the utmost degree ... I do not know what Ts'ao Kung makes these Nine Variations out to be [the latter's note is 變其正得其所用九也], but it has been suggested that they are connected with the Nine Situations" — of chap. XI. This is the view adopted by Chang Yü: see note on 死地, § 2. The only other alternative is to suppose that something has been lost — a supposition to which the unusual shortness of the chapter lends some weight.

1. Sun Tzŭ said: In war, the general receives his commands from the sovereign, collects his army and concentrates his forces.

Repeated from VII. § 1, where it is certainly more in place. It may have been interpolated here merely in order to supply a beginning to the chapter.

2. When in difficult country, do not encamp.

For explanation of 圮地, see XI. § 8.

In country where high roads intersect, join hands with your allies.

See XI, §§ 6, 12. Capt. Calthrop omits 衢地.

Do not linger in dangerously isolated positions.

絕 地 is not one of the Nine Situations as given in the beginning of chap. XI, but occurs later on (*ibid.* § 43, *q. v.*). We may compare it with 重 地 (XI. §7). Chang Yü calls it a 危 絕 之 地, situated across the frontier, in hostile territory. Li Ch'üan says it is "country in which there are no springs or wells, flocks or herds, vegetables or firewood;" Chia Lin, "one of gorges, chasms and precipices, without a road by which to advance."

In hemmed-in situations, you must resort to stratagem.

See XI. §§ 9, 14. Capt. Calthrop has "mountainous and wooded country," which is a quite inadequate translation of 圍.

In a desperate position, you must fight.

See XI. §§ 10, 14. Chang Yü has an important note here, which must be given in full. "From 圮 地 無 舍," he says, "down to this point, the Nine Variations are presented to us. The reason why only five are given is that the subject is treated *en précis* (舉 其 大 略 也). So in chap. XI, where he discusses the variations of tactics corresponding to the Nine Grounds, Sun Tzǔ mentions only six variations; there again we have an abridgment. [I cannot understand what Chang Yü means by this statement. He can only be referring to §§ 11—14 or §§ 46—50 of chap. XI; but in both places all the nine grounds are discussed. Perhaps he is confusing these with the Six 地 形 of chap. X.] All kinds of ground have corresponding military positions, and also a variation of tactics suitable to each (凡 地 有 勢 有 變). In chap. XI, what we find enumerated first [§§ 2—10] are the situations; afterwards [§§ 11—14] the corresponding tactics. Now, how can we tell that the 九 變 "Nine Variations" are simply the 九 地 之 變 "variations of tactics corresponding to the Nine Grounds"? It is said further on [§ 5] that 'the general who does not understand the nine variations of tactics may be well acquainted with the features of the country, yet he will not be able to turn his knowledge to practical account.' Again, in chap. XI [§ 41] we read: 'The different measures adapted to the nine varieties of ground (九 地 之 變) and the expediency of aggressive or defensive tactics must be carefully examined.' From a consideration of these passages the meaning is made clear. When later on the nine grounds are enumerated, Sun Tzǔ recurs to these nine variations. He wishes here to speak of the Five Advantages [see *infra*, § 6], so he begins by setting forth the Nine Variations. These are inseparably connected in practice, and therefore they are dealt with together." The weak point of this argument is the suggestion that 五 事 "five things" can stand as a 大 畧, that is, an

3. 塗有所不由軍有所不擊城有所不攻地有
所不爭君命有所不受

abstract or abridgment, of nine, when those that are omitted are not less important than those that appear, and when one of the latter is not included amongst the nine at all.

3. **There are roads which must not be followed,**

"Especially those leading through narrow defiles," says Li Ch'üan, "where an ambush is to be feared."

armies which must not be attacked,

More correctly, perhaps, "there are times when an army must not be attacked." Ch'ên Hao says: "When you see your way to obtain a trivial advantage, but are powerless to inflict a real defeat, refrain from attacking, for fear of overtaxing your men's strength."

towns

Capt. Calthrop says "castles" — an unfortunate attempt to introduce local colour.

which must not be besieged,

Cf. III. § 4. Ts'ao Kung gives an interesting illustration from his own experience. When invading the territory of 徐州 Hsü-chou, he ignored the city of 華費 Hua-pi, which lay directly in his path, and pressed on into the heart of the country. This excellent strategy was rewarded by the subsequent capture of no fewer than fourteen important district cities. Chang Yü says: "No town should be attacked which, if taken, cannot be held, or if left alone, will not cause any trouble" 荀罃 Hsün Ying, when urged to attack 偪陽 Pi-yang, replied: "The city is small and well-fortified; even if I succeed in taking it, 't will be no great feat of arms; whereas if I fail, I shall make myself a laughing-stock." In the seventeenth century, sieges still formed a large proportion of war. It was Turenne who directed attention to the importance of marches, countermarches and manœuvres. He said: "It is a great mistake to waste men in taking a town when the same expenditure of soldiers will gain a province." *

positions which must not be contested, commands of the sovereign which must not be obeyed.

This is a hard saying for the Chinese, with their reverence for authority, and Wei Liao Tzŭ (quoted by Tu Mu) is moved to exclaim:

---

* "Marshal Turenne," p. 50.

4. 故將通於九變之利者知用兵矣
5. 將不通於九變之利者雖知地形不能得地
之利矣
6. 治兵不知九變之術雖知五利不能得人之
用矣

兵者凶器也爭者逆德也將者死官也 "Weapons
are baleful instruments, strife is antagonistic to virtue, a military commander is the negation of civil order!" The unpalatable fact remains,
however, that even Imperial wishes must be subordinated to military
necessity. Cf. III. § 17. (5), X. § 23. The *T'ung Tien* has 將在軍
before 君命, etc. This is a gloss on the words by Chu-ko Liang, which
being repeated by Tu Yu became incorporated with the text. Chang Yü
thinks that these five precepts are the 五利 referred to in § 6. Another
theory is that the mysterious 九變 are here enumerated, starting with

圮地無舍 and ending at 地有所不爭, while the final clause

君命有所不受 embraces and as it were sums up all the nine. Thus
Ho Shih says: "Even if it be your sovereign's command to encamp in difficult country, linger in isolated positions, etc., you must not do so." The
theory is perhaps a little too ingenious to be accepted with confidence.

4. The general who thoroughly understands the advantages that accompany variation of tactics knows how to
handle his troops.

Before 利 in the original text there is a 地 which is obviously not
required.

5. The general who does not understand these, may be
well acquainted with the configuration of the country, yet he
will not be able to turn his knowledge to practical account.

Literally, "get the advantage of the ground," which means not only
securing good positions, but availing oneself of natural advantages in
every possible way. Chang Yü says: "Every kind of ground is characterised
by certain natural features, and also gives scope for a certain variability
of plan. How is it possible to turn these natural features to account
unless topographical knowledge is supplemented by versatility of mind?"

6. So, the student of war who is unversed in the art
of varying his plans, even though he be acquainted with
the Five Advantages, will fail to make the best use of
his men.

7. 是故智者之慮必雜於利害
8. 雜於利而務可信也
9. 雜於害而患可解也

Ts'ao Kung says that the 五利 are 下五事也 "the five things that follow;" but this cannot be right. We must rather look back to the five "variations" contained in § 3. Chia Lin (who reads 五變 here to balance the 五利) tells us that these imply five obvious and generally advantageous lines of action, namely: "if a certain road is short, it must be followed; if an army is isolated, it must be attacked; if a town is in a parlous condition, it must be besieged; if a position can be stormed, it must be attempted; and if consistent with military operations, the ruler's commands must be obeyed." But there are circumstances which sometimes forbid a general to use these advantages. For instance, "a certain road may be the shortest way for him, but if he knows that it abounds in natural obstacles, or that the enemy has laid an ambush on it, he will not follow that road. A hostile force may be open to attack, but if he knows that it is hard-pressed and likely to fight with desperation, he will refrain from striking," and so on. Here the 變 comes in to modify the 利, and hence we see the uselessness of knowing the one without the other — of having an eye for weaknesses in the enemy's armour without being clever enough to recast one's plans on the spur of the moment. Capt. Calthrop offers this slovenly translation: "In the management of armies, if the art of the Nine Changes be understood [*sic*], a knowledge of the Five Advantages is of no avail."

7. Hence in the wise leader's plans, considerations of advantage and of disadvantage will be blended together.

"Whether in an advantageous position or a disadvantageous one," says Ts'ao Kung, "the opposite state should be always present to your mind."

8. If our expectation of advantage be tempered in this way, we may succeed in accomplishing the essential part of our schemes.

信, according to Tu Mu, is equivalent to 申, and 務可信也 is paraphrased by Chang Yü as 可以伸己之事. Tu Mu goes on to say: "If we wish to wrest an advantage from the enemy, we must not fix our minds on that alone, but allow for the possibility of the enemy also doing some harm to us, and let this enter as a factor into our calculations."

9. If, on the other hand, in the midst of difficulties we

10. 是故屈諸侯者以害役諸侯者以業趨諸侯者以利

are always ready to seize an advantage, we may extricate ourselves from misfortune.

A translator cannot emulate the conciseness of 雜於害 "to blend [thoughts of advantage] with disadvantage," but the meaning is as given. Tu Mu says: "If I wish to extricate myself from a dangerous position, I must consider not only the enemy's ability to injure me, but also my own ability to gain an advantage over the enemy. If in my counsels these two considerations are properly blended, I shall succeed in liberating myself... For instance, if I am surrounded by the enemy and only think of effecting an escape, the nervelessness of my policy will incite my adversary to pursue and crush me; it would be far better to encourage my men to deliver a bold counter-attack, and use the advantage thus gained to free myself from the enemy's toils." See the story of Ts'ao Ts'ao, VII. § 35, note. In his first edition, Capt. Calthrop translated §§ 7—9 as follows: "The wise man perceives clearly wherein lies advantage and disadvantage. While recognising an opportunity, he does not overlook the risks, and saves future anxiety." This has now been altered into: "The wise man considers well both advantage and disadvantage. He sees a way out of adversity, *and on the day of victory to danger is not blind.*" Owing to a needless inversion of the Chinese, the words which I have italicised are evidently intended to represent § 8!

10. Reduce the hostile chiefs by inflicting damage on them;

Chia Lin enumerates several ways of inflicting this injury, some of which would only occur to the Oriental mind: — "Entice away the enemy's best and wisest men, so that he may be left without counsellors. Introduce traitors into his country, that the government policy may be rendered futile. Foment intrigue and deceit, and thus sow dissension between the ruler and his ministers. By means of every artful contrivance, cause deterioration amongst his men and waste of his treasure. Corrupt his morals by insidious gifts leading him into excess. Disturb and unsettle his mind by presenting him with lovely women." Chang Yü (after Wang Hsi) considers the 害 to be military chastisement: "Get the enemy," he says, "into a position where he must suffer injury, and he will submit of his own accord." Capt. Calthrop twists Sun Tzŭ's words into an absurdly barbarous precept: "In reducing an enemy to submission, inflict all possible damage upon him."

make trouble for them,

業 is defined by Ts'ao Kung as 事, and his definition is generally

11. 故用兵之法無恃其不來恃吾有以待也無恃其不攻恃吾有所不可攻也

12. 故將有五危必死可殺也必生可虜也忿速可侮也廉潔可辱也愛民可煩也

adopted by the commentators. Tu Mu, however, seems to take it in the sense of "possessions," or, as we might say, "assets," which he considers to be 兵眾國富人和令行 "a large army, a rich exchequer, harmony amongst the soldiers, punctual fulfilment of commands." These give us a whip-hand over the enemy.

and keep them constantly engaged;

役, literally, "make servants of them." Tu Yu says 令不得安佚 "prevent them from having any rest."

hold out specious allurements, and make them rush to any given point.

Mêng Shih's note contains an excellent example of the idiomatic use of 變: 令忘變而速至 "cause them to forget *pien* (the reasons for acting otherwise than on their first impulse), and hasten in our direction."

11. The art of war teaches us to rely not on the likelihood of the enemy's not coming, but on our own readiness to receive him;

The *T'ung Tien* and *Yü Lan* read 有能以待之也, but the conciser form is more likely to be right.

not on the chance of his not attacking, but rather on the fact that we have made our position unassailable.

The *T'ung Tien* and *Yü Lan* insert 吾也 after the first 攻, and omit 有所.

12. There are five dangerous faults which may affect a general: (1) Recklessness, which leads to destruction;

勇而無慮 "Bravery without forethought," as Ts'ao Kung analyses it, which causes a man to fight blindly and desperately like a mad bull. Such an opponent, says Chang Yü, "must not be encountered with brute force, but may be lured into an ambush and slain." Cf. Wu Tzŭ, chap. IV *ad init.*: 凡人論將常觀於勇勇之於將乃數分

之 一 耳 夫 勇 者 必 輕 合 輕 合 而 不 知 利 未 可 也

"In estimating the character of a general, men are wont to pay exclusive attention to his courage, forgetting that courage is only one out of many qualities which a general should possess. The merely brave man is prone to fight recklessly; and he who fights recklessly, without any perception of what is expedient, must be condemned." Ssŭ-ma Fa, too, makes the incisive remark 上 死 不 勝 "Simply going to one's death does not bring about victory."

(2) cowardice, which leads to capture;

必 生 is explained by Ts'ao Kung of the man "whom timidity prevents from advancing to seize an advantage," and Wang Hsi adds, "who is quick to flee at the sight of danger." Mêng Shih gives the closer paraphrase 志 必 生 反 "he who is bent on returning alive," that is, the man who will never take a risk. But, as Sun Tzŭ knew, nothing is to be achieved in war unless you are willing to take risks. T'ai Kung said: 失 利 後 時 反 受 其 殃 "He who lets an advantage slip will subsequently bring upon himself real disaster." In 404 A.D., 劉 裕 Liu Yü pursued the rebel 桓 玄 Huan Hsüan up the Yangtsze and fought a naval battle with him at 崢 嶸 洲 the island of Ch'êng-hung. The loyal troops numbered only a few thousands, while their opponents were in great force. But Huan Hsüan, fearing the fate which was in store for him should he be overcome, had a light boat made fast to the side of his war-junk, so that he might escape, if necessary, at a moment's notice. The natural result was that the fighting spirit of his soldiers was utterly quenched, and when the loyalists made an attack from windward with fireships, all striving with the utmost ardour to be first in the fray, Huan Hsüan's forces were routed, had to burn all their baggage and fled for two days and nights without stopping. [See 晉 書, chap. 99, fol. 13.] Chang Yü tells a somewhat similar story of 趙 嬰 齊 Chao Ying-ch'i, a general of the Chin State who during a battle with the army of Ch'u in 597 B.C. had a boat kept in readiness for him on the river, wishing in case of defeat to be the first to get across.

(3) a hasty temper, which can be provoked by insults;

I fail to see the meaning of Capt. Calthrop's "which *brings* insult." Tu Mu tells us that 姚 襄 Yao Hsiang, when opposed in 357 A.D. by 黃 眉 Huang Mei, 鄧 羌 Têng Ch'iang and others, shut himself up behind his walls and refused to fight. Têng Ch'iang said: "Our adversary is of a choleric temper and easily provoked; let us make constant sallies and break down his walls, then he will grow angry and come out.

13. 凡此五者將之過也用兵之災也
14. 覆軍殺將必以五危不可不察也

Once we can bring his force to battle, it is doomed to be our prey." This plan was acted upon, Yao Hsiang came out to fight, was lured on as far as 三原 San-yüan by the enemy's pretended flight, and finally attacked and slain.

(4) a delicacy of honour which is sensitive to shame;

This need not be taken to mean that a sense of honour is really a defect in a general. What Sun Tzŭ condemns is rather an exaggerated sensitiveness to slanderous reports, the thin-skinned man who is stung by opprobrium, however undeserved. Mei Yao-ch'ên truly observes, though somewhat paradoxically: 徇名不顧 "The seeker after glory should be careless of public opinion."

(5) over-solicitude for his men, which exposes him to worry and trouble.

Here again, Sun Tzŭ does not mean that the general is to be careless of the welfare of his troops. All he wishes to emphasise is the danger of sacrificing any important military advantage to the immediate comfort of his men. This is a shortsighted policy, because in the long run the troops will suffer more from the defeat, or, at best, the prolongation of the war, which will be the consequence. A mistaken feeling of pity will often induce a general to relieve a beleaguered city, or to reinforce a hard-pressed detachment, contrary to his military instincts. It is now generally admitted that our repeated efforts to relieve Ladysmith in the South African War were so many strategical blunders which defeated their own purpose. And in the end, relief came through the very man who started out with the distinct resolve no longer to subordinate the interests of the whole to sentiment in favour of a part. An old soldier of one of our generals who failed most conspicuously in this war, tried once, I remember, to defend him to me on the ground that he was always "so good to his men." By this plea, had he but known it, he was only condemning him out of Sun Tzŭ's mouth.

13. These are the five besetting sins of a general, ruinous to the conduct of war.

14. When an army is overthrown and its leader slain, the cause will surely be found among these five dangerous faults. Let them be a subject of meditation.

1. 孫子曰凡處軍相敵絕山依谷

IX. THE ARMY ON THE MARCH.

The contents of this interesting chapter are better indicated in § 1 than by this heading.

1. Sun Tzŭ said: We come now to the question of encamping the army, and observing signs of the enemy.

The discussion of 處軍, as Chang Yü points out, extends from here down to 伏姦之所藏處也 (§§ 1—17), and 相敵 from that point down to 必謹察之 (§§ 18—39). The rest of the chapter consists of a few desultory remarks, chiefly on the subject of discipline.

Pass quickly over mountains,

For this use of 絕, cf. *infra*, § 3. See also 荀子, ch. 1. fol. 2 (standard edition of 1876): 絕江河; *Shih Chi*, ch. 27 *ad init.*: 後六星絕漢.

and keep in the neighbourhood of valleys.

Tu Mu says that 依 here = 近. The idea is, not to linger among barren uplands, but to keep close to supplies of water and grass. Capt. Calthrop translates "camp in valleys," heedless of the very next sentence. Cf. Wu Tzŭ, ch. 3: 無當天竈 "Abide not in natural ovens," *i.e.* 大谷之口 "the openings of large valleys." Chang Yü tells the following anecdote: "武都羌 Wu-tu Ch'iang was a robber captain in the time of the Later Han, and 馬援 Ma Yüan was sent to exterminate his gang. Ch'iang having found a refuge in the hills, Ma Yüan made no attempt to force a battle, but seized all the favourable positions commanding supplies of water and forage. Ch'iang was soon in such a desperate plight for want of provisions that he was forced to make a total surrender. He did not know the advantage of keeping in the neighbourhood of valleys."

2. 視生處高戰隆無登此處山之軍也
3. 絶水必遠水
4. 客絶水而來勿迎之於水內令半濟而擊之
   利

2. Camp in high places,

Not on high hills, but on knolls or hillocks elevated above the surrounding country.

facing the sun.

視生＝面陽. Tu Mu takes this to mean "facing south," and Ch'ên Hao "facing east." Cf. *infra*, §§ 11, 13.

Do not climb heights in order to fight.

隆 is here simply equivalent to 高. The *T'ung Tien* and *Yü Lan* read 降.

So much for mountain warfare.

After 山, the *T'ung Tien* and *Yü Lan* insert 谷.

3. After crossing a river, you should get far away from it.

"In order to tempt the enemy to cross after you," according to Ts'ao Kung, and also, says Chang Yü, "in order not to be impeded in your evolutions." · The *T'ung Tien* reads 敵若絶水 "If *the enemy* crosses a river," etc. But in view of the next sentence, this is almost certainly an interpolation.

4. When an invading force crosses a river in its onward march, do not advance to meet it in mid-stream. It will be best to let half the army get across, and then deliver your attack.

The *T'ung Tien* and *Yü Lan* read 度 for 濟, without change of meaning. Wu Tzŭ plagiarises this passage twice over: — ch. II *ad fin.*, 涉水半渡可擊; ch. V, 敵若絶水半渡而擊. Li Ch'üan alludes to the great victory won by Han Hsin over 龍且 Lung Chü at the 濰 Wei River. Turning to the *Ch'ien Han Shu*, ch. 34, fol. 6 *verso*, we find the battle described as follows: "The two armies were drawn up on opposite sides of the river. In the night, Han Hsin ordered his men to take some ten thousand sacks filled with sand and construct a dam a little higher up. Then, leading half his army across, he at-

5. 欲戰者無附於水而迎客
6. 視生處高無迎水流此處水上之軍也

tacked Lung Chü; but after a time, pretending to have failed in his attempt, he hastily withdrew to the other bank. Lung Chü was much elated by this unlooked-for success, and exclaiming: "I felt sure that Han Hsin was really a coward!" he pursued him and began crossing the river in his turn. Han Hsin now sent a party to cut open the sandbags, thus releasing a great volume of water, which swept down and prevented the greater portion of Lung Chü's army from getting across. He then turned upon the force which had been cut off, and annihilated it, Lung Chü himself being amongst the slain. The rest of the army, on the further bank, also scattered and fled in all directions."

5. If you are anxious to fight, you should not go to meet the invader near a river which he has to cross.

For fear of preventing his crossing. Capt. Calthrop makes the injunction ridiculous by omitting 欲戰者.

6. Moor your craft higher up than the enemy, and facing the sun.

See *supra*, § 2. The repetition of these words in connection with water is very awkward. Chang Yü has the note: 或岸邊爲陳或水上泊舟皆須面陽而居高 "Said either of troops marshalled on the river-bank, or of boats anchored in the stream itself; in either case it is essential to be higher than the enemy and facing the sun." The other commentators are not at all explicit. One is much tempted to reject their explanation of 視生 altogether, and understand it simply as "seeking safety." [Cf. 必生 in VIII. § 12, and *infra*, § 9.] It is true that this involves taking 視 in an unusual, though not, I think, an impossible sense. Of course the earlier passage would then have to be translated in like manner.

Do not move up-stream to meet the enemy.

Tu Mu says: "As water flows downwards, we must not pitch our camp on the lower reaches of a river, for fear the enemy should open the sluices and sweep us away in a flood. This is implied above in the words 視生處高. Chu-ko Wu-hou has remarked that 'in river warfare we must not advance against the stream,' which is as much as to say that our fleet must not be anchored below that of the enemy, for then they would be able to take advantage of the current and make short work of us." There is also the danger, noted by other commentators,

7. 絕斥澤惟亟去無留

8. 若交軍於斥澤之中必依水草而背眾樹此
處斥澤之軍也

9. 平陸處易而右背高前死後生此處平陸之
軍也

that the enemy may throw poison on the water to be carried down to us. Capt. Calthrop's first version was: "Do not cross rivers in the face of the stream" — a sapient piece of advice, which made one curious to know what the correct way of crossing rivers might be. He has now improved this into: "Do not fight when the enemy is between the army and the source of the river."

So much for river warfare.

7. In crossing salt-marshes, your sole concern should be to get over them quickly, without any delay.

Because of the lack of fresh water, the poor quality of the herbage, and last but not least, because they are low, flat, and exposed to attack.

8. If forced to fight in a salt-marsh, you should have water and grass near you, and get your back to a clump of trees.

Li Ch'üan remarks that the ground is less likely to be treacherous where there are trees, while Tu Yu says that they will serve to protect the rear. Capt. Calthrop, with a perfect genius for going wrong, says "in the neighbourhood of a marsh." For 若 the *T'ung Tien* and *Yü Lan* wrongly read 為, and the latter also has 倍 instead of 背.

So much for operations in salt-marshes.

9. In dry, level country, take up an easily accessible position

This is doubtless the force of 易, its opposite being 險. Thus, Tu Mu explains it as 坦易平穩之處 "ground that is smooth and firm," and therefore adapted for cavalry; Chang Yü as 坦易無坎陷之處 "level ground, free from depressions and hollows." He adds later on that although Sun Tzŭ is discussing flat country, there will nevertheless be slight elevations and hillocks.

with rising ground to your right and on your rear,

10. 凡此四軍之利黃帝之所以勝四帝也
11. 凡軍喜高而惡下貴陽而賤陰
12. 養生而處實軍無百疾是謂必勝

The Yü Lan again reads 倍 for 背. Tu Mu quotes T'ai Kung as saying: "An army should have a stream or a marsh on its left, and a hill or tumulus on its right."

so that the danger may be in front, and safety lie behind.

Wang Hsi thinks that 後生 contradicts the saying 視生 in § 2, and therefore suspects a mistake in the text.

So much for campaigning in flat country.

10. These are the four useful branches of military knowledge

Those, namely, concerned with (1) mountains, (2) rivers, (3) marshes, and (4) plains. Compare Napoleon's "Military Maxims," no. 1.

which enabled the Yellow Emperor to vanquish four several sovereigns.

Mei Yao-ch'ên asks, with some plausibility, whether 帝 is not a mistake for 軍 "armies," as nothing is known of Huang Ti having conquered four other Emperors. The Shih Chi (ch. I ad init.) speaks only of his victories over 炎帝 Yen Ti and 蚩尤 Ch'ih Yu. In the 六韜 it is mentioned that he "fought seventy battles and pacified the Empire." Ts'ao Kung's explanation is, that the Yellow Emperor was the first to institute the feudal system of vassal princes, each of whom (to the number of four) originally bore the title of Emperor. Li Ch'üan tells us that the art of war originated under Huang Ti, who received it from his Minister 風后 Fêng Hou.

11. All armies prefer high ground to low,

"High ground," says Mei Yao-ch'ên, "is not only more agreeable and salubrious, but more convenient from a military point of view; low ground is not only damp and unhealthy, but also disadvantageous for fighting." The original text and the T'u Shu have 好 instead of 喜.

and sunny places to dark.

12. If you are careful of your men,

Ts'ao Kung says: 向水草可放牧養畜 "Make for fresh water and pasture, where you can turn out your animals to graze." And

13. 邱陵隄防必處其陽而右背之此兵之利地
之助也

14. 上雨水沫至欲涉者待其定也

15. 凡地有絕澗天井天牢天羅天陷天隙必亟
去之勿近也

the other commentators follow him, apparently taking 生 as = 性.
Cf. Mencius, V. i. ix. i, where 養牲者 means a cattle-keeper. But
here 養生 surely has reference to the health of the troops. It is the
title of Chuang Tzŭ's third chapter, where it denotes moral rather than
physical well-being.

and camp on hard ground,

實 must mean dry and solid, as opposed to damp and marshy, ground.
This is to be found as a rule in high places, so the commentators explain
實 as practically equivalent to 高.

the army will be free from disease of every kind,

Chang Yü says: "The dryness of the climate will prevent the outbreak
of illness."

and this will spell victory.

13. When you come to a hill or a bank, occupy the
sunny side, with the slope on your right rear. Thus you
will at once act for the benefit of your soldiers and utilise
the natural advantages of the ground.

14. When, in consequence of heavy rains up-country,
a river which you wish to ford is swollen and flecked
with foam, you must wait until it subsides.

The *T'ung Tien* and *Yü Lan* have a superfluous 下 before 水.

15. Country in which there are precipitous cliffs with
torrents running between,

絕澗, explained by Mei Yao-ch'ên as 前後險峻水橫其中.
deep natural hollows,

天井, explained as 四面峻坂澗壑所歸 "places enclosed
on every side by steep banks, with pools of water at the bottom."

confined places,

16. 吾遠之敵近之吾迎之敵背之

17. 軍旁有險阻蔣潢井生葭葦小林蘙薈必謹
覆索之此伏姦之所藏處也

天牢 "natural pens or prisons," explained as 三面環絕易
入難出 "places surrounded by precipices on three sides — easy to
get into, but hard to get out of."

tangled thickets,

天羅, explained as 草木蒙密鋒鏑莫施 "places covered
with such dense undergrowth that spears cannot be used."

quagmires

天陷, explained as 卑下汙濘車騎不通 "low-lying places,
so heavy with mud as to be impassible for chariots and horsemen."

and crevasses,

天隙 is explained by Mei Yao-ch'ên as 兩山相向洞道狹惡
"a narrow difficult way between beetling cliffs," but Ts'ao Kung says
山澗迫狹地形深數尺長數丈者, which seems to
denote something on a much smaller scale. Tu Mu's note is 地多溝
坑坎陷木石 "ground covered with trees and rocks, and inter-
sected by numerous ravines and pitfalls." This is very vague, but Chia
Lin explains it clearly enough as a defile or narrow pass: 兩邊險絕
形狹長而數里, and Chang Yü takes much the same view.
On the whole, the weight of the commentators certainly inclines to the
rendering "defile". But the ordinary meaning of 隙 (a crack or fissure)
and the fact that 絕澗 above must be something in the nature of a
defile, make me think that Sun Tzŭ is here speaking of crevasses. The
T'ung Tien and *Yü Lan* read 郄 for 隙, with the same meaning; the
latter also has 大害 after 天郄 — a palpable gloss.

should be left with all possible speed and not approached

16. While we keep away from such places, we should
get the enemy to approach them; while we face them,
we should let the enemy have them on his rear.

17. If in the neighbourhood of your camp

The original text has 軍行, but 旁 has been generally adopted as
yielding much better sense.

18. 敵近而靜者恃其險也

there should be any hilly country,

險阻 is 邱阜之地, according to Chang Yü.

ponds surrounded by aquatic grass, hollow basins filled with reeds,

The original text omits 蔣 and 生, so that 潢 and 井 join to make a pair: "ponds and basins." This is plausible enough at first sight, but there are several objections to the reading: (1) 蔣 is unlikely to have got into the text as a gloss on 潢; (2) it is easy to suppose, on the other hand, that 蔣 and afterwards 生 (to restore the balance of the sentence) were omitted by a copyist who jumped to the conclusion that 潢 and 井 must go together; (3) the sense, when one comes to consider it, actually requires 蔣, for it is absurd to talk of pools and ponds as in themselves suitable places for an ambush; (4) Li Ching (571—649 A.D.) in his 兵法 "Art of War" has the words: 蔣潢翳薈則必索其伏. This is evidently a reminiscence of Sun Tzŭ, so there can be little doubt that 蔣 stood in the text at this early date. It may be added that the *T'ung Tien* and *Yu Lan* both have 蔣, and the latter also reads 并 for 井.

or woods with thick undergrowth,

I read 小林 with the *Yü Lan* in preference to 山林, given in the original text, which is accepted by the commentators without question. The text of the *T'u Shu* up to this point runs as follows: 潢井蒹葭林木翳薈者.

they must be carefully routed out and searched; for these are places where men in ambush or insidious spies are likely to be lurking.

The original text omits 藏, which has been restored from the *T'ung Tien* and *Yü Lan*. The *T'u Shu* omits 處 as well, making 所 a substantive. On 姦 Chang Yü has the note: 又慮姦細潛隱覘我虛實聽我號令伏姦當爲兩事 "We must also be on our guard against traitors who may lie in close covert, secretly spying out our weaknesses and overhearing our instructions. *Fu* and *chien* are to be taken separately."

18. When the enemy is close at hand and remains quiet, he is relying on the natural strength of his position.

19. 遠而挑戰者欲人之進也
20. 其所居者易利也
21. 衆樹動者來也衆草多障者疑也

Here begin Sun Tzŭ's remarks on the reading of signs, much of which is so good that it could almost be included in a modern manual like Gen. Baden-Powell's "Aids to Scouting."

19. When he keeps aloof and tries to provoke a battle, he is anxious for the other side to advance.

Probably because we are in a strong position from which he wishes to dislodge us. "If he came close up to us," says Tu Mu, "and tried to force a battle, he would seem to despise us, and there would be less probability of our responding to the challenge."

20. If his place of encampment is easy of access, he is tendering a bait.

易 is here the opposite of 險 in § 18. The reading of the *T'ung Tien* and *Yü Lan*, 其所處者居易利也, is pretty obviously corrupt. The original text, which transposes 易 and 者, may very possibly be right. Tu Mu tells us that there is yet another reading: 士爭其所居者易利也.

21. Movement amongst the trees of a forest shows that the enemy is advancing.

Ts'ao Kung explains this as "felling trees to clear a passage," and Chang Yü says: "Every army sends out scouts to climb high places and observe the enemy. If a scout sees that the trees of a forest are moving and shaking, he may know that they are being cut down to clear a passage for the enemy's march."

The appearance of a number of screens in the midst of thick grass means that the enemy wants to make us suspicious.

Whenever the meaning of a passage happens to be somewhat elusive, Capt. Calthrop seems to consider himself justified in giving free rein to the imagination. Thus, though his text is here identical with ours, he renders the above: "Broken branches and trodden grass, as of the passing of a large host, must be regarded with suspicion." Tu Yu's explanation, borrowed from Ts'ao Kung, is as follows: "The presence of a number of screens or sheds in the midst of thick vegetation is a sure sign that the enemy has fled and, fearing pursuit, has constructed these hiding-places

22. 鳥起者伏也獸駭者覆也

23. 塵高而銳者車來也卑而廣者徒來也散而
條達者樵採也少而往來者營軍也

in order to make us suspect an ambush." It appears that these "screens" were hastily knotted together out of any long grass which the retreating enemy happened to come across.

22. The rising of birds in their flight is the sign of an ambuscade.

Chang Yü's explanation is doubtless right: "When birds that are flying along in a straight line suddenly shoot upwards, it means that soldiers are in ambush at the spot beneath."

Startled beasts indicate that a sudden attack is coming.

An example of 覆 *fou*[4] in the meaning of "ambuscade" may be found in the *Tso Chuan*, 隱 9th year: 君爲三覆以待之. In the present passage, however, it is to be distinguished from 伏 just above, in that it implies onward motion on the part of the attacking force. Thus, Li Ch'üan defines it as 不意而至, and Tu Mu as 來襲我也.

23. When there is dust rising in a high column, it is the sign of chariots advancing; when the dust is low, but spread over a wide area, it betokens the approach of infantry.

高而銳 "high and sharp," or rising to a peak, is of course somewhat exaggerated as applied to dust. The commentators explain the phenomenon by saying that horses and chariots, being heavier than men, raise more dust, and also follow one another in the same wheel-track, whereas foot-soldiers would be marching in ranks, many abreast. According to Chang Yü, "every army on the march must have scouts (探候之人) some way in advance, who on sighting dust raised by the enemy, will gallop back and report it to the commander-in-chief." Cf. Gen. Baden-Powell: "As you move along, say, in a hostile country, your eyes should be looking afar for the enemy or any signs of him: figures, dust rising, birds getting up, glitter of arms, etc." *

When it branches out in different directions, it shows that parties have been sent to collect firewood.

There is some doubt about the reading 樵採. The *T'ung Tien* and *Yu Lan* have 薪採, and Li Ch'üan proposes 薪來.

---

* "Aids to Scouting," p. 26.

24. 辭卑而益備者進也辭强而進驅者退也

A few clouds of dust moving to and fro signify that the army is encamping.

Chang Yü says: "In apportioning the defences for a cantonment, light horse will be sent out to survey the position and ascertain the weak and strong points all along its circumference. Hence the small quantity of dust and its motion."

24. Humble words and increased. preparations are signs that the enemy is about to advance.

"As though they stood in great fear of us," says Tu Mu. "Their object is to make us contemptuous and careless, after which they will attack us." Chang Yü alludes to the story of 田單 T'ien Tan of the Ch'i State, who in 279 B.C. was hard-pressed in his defence of 即墨 Chi-mo against the Yen forces, led by 騎劫 Ch'i Chieh. In ch. 82 of the *Shih Chi* we read: "T'ien Tan openly said: 'My only fear is that the Yen army may cut off the noses of their Ch'i prisoners and place them in the front rank to fight against us; that would be the undoing of our city.' The other side being informed of this speech, at once acted on the suggestion; but those within the city were enraged at seeing their fellow-countrymen thus mutilated, and fearing only lest they should fall into the enemy's hands, were nerved to defend themselves more obstinately than ever. Once again T'ien Tan sent back converted spies who reported these words to the enemy: 'What I dread most is that the men of Yen may dig up the ancestral tombs outside the town, and by inflicting this indignity on our forefathers cause us to become faint-hearted.' Forthwith the besiegers dug up all the graves and burned the corpses lying in them. And the inhabitants of Chi-mo, witnessing the outrage from the city-walls, wept passionately and were all impatient to go out and fight, their fury being increased tenfold. T'ien Tan knew then that his soldiers were ready for any enterprise. But instead of a sword, he himself took a mattock in his hands, and ordered others to be distributed amongst his best warriors, while the ranks were filled up with their wives and concubines. He then served out all the remaining rations and bade his men eat their fill. The regular soldiers were told to keep out of sight, and the walls were manned with the old and weaker men and with women. This done, envoys were despatched to the enemy's camp to arrange terms of surrender, whereupon the Yen army began shouting for joy. T'ien Tan also collected 20,000 ounces of silver from the people, and got the wealthy citizens of Chi-mo to send it to the Yen general with the prayer that, when the town capitulated, he would not allow their homes to be plundered or their women to be maltreated. Ch'i Chieh, in high good humour, granted their prayer; but his army now became increasingly slack and

25. 輕車先出居其側者陳也

26. 無約而請和者謀也

careless. Meanwhile, T'ien Tan got together a thousand oxen, decked them with pieces of red silk, painted their bodies, dragon-like, with coloured stripes, and fastened sharp blades on their horns and well-greased rushes on their tails. When night came on, he lighted the ends of the rushes, and drove the oxen through a number of holes which he had pierced in the walls, backing them up with a force of 5000 picked warriors. The animals, maddened with pain, dashed furiously into the enemy's camp where they caused the utmost confusion and dismay; for their tails acted as torches, showing up the hideous pattern on their bodies, and the weapons on their horns killed or wounded any with whom they came into contact. In the meantime, the band of 5000 had crept up with gags in their mouths, and now threw themselves on the enemy. At the same moment a frightful din arose in the city itself, all those that remained behind making as much noise as possible by banging drums and hammering on bronze vessels, until heaven and earth were convulsed by the uproar. Terror-stricken, the Yen army fled in disorder, hotly pursued by the men of Ch'i, who succeeded in slaying their general Ch'i Chieh... The result of the battle was the ultimate recovery of some seventy cities which had belonged to the Ch'i State."

Violent language and driving forward as if to the attack are signs that he will retreat.

I follow the original text here, also adopted by the *T'u Shu*. The standard text reads 辭詭而強進驅者退也 on the strength of Ts'ao Kung's commentary 詭詐也, which shows that his text included the word 詭. Strong as this ground is, I do not think it can counterbalance the obvious superiority of the other reading in point of sense. 詭 not only provides no antithesis to 卑, but makes the whole passage absurd; for if the language of the enemy is calculated to deceive, it cannot be known as deceitful at the time, and can therefore afford no "sign." Moreover, the extra word in 強進驅者 (an awkward locution, by the way) spoils the parallelism with 益備者.

25. When the light chariots

The same, according to Tu Yu, as the 馳車 of II. § 1.

come out first and take up a position on the wings, it is a sign that the enemy is forming for battle.

The *T'ung Tien* omits 出.

26. Peace proposals unaccompanied by a sworn covenant indicate a plot.

27. 奔走而陳兵者期也
28. 半進半退者誘也
29. 倚仗而立者飢也
30. 汲而先飲者渴也

Tu Yu defines 約 as 要約, and Li Ch'üan as 質盟之約 "a treaty confirmed by oaths and hostages." Wang Hsi and Chang Yü, on the other hand, simply say 無故 "without reason," "on a frivolous pretext," as though 約 bore the rather unusual sense of "important." Capt. Calthrop has "without consultation," which is too loose.

27. **When there is much running about**

Every man hastening to his proper place under his own regimental banner.

and the soldiers fall into rank,

I follow the *T'u Shu* in omitting 車 after 兵. Tu Mu quotes the *Chou Li*, ch. xxix. fol. 31: 車驟徒趨及表乃止.

it means that the critical moment has come.

What Chia Lin calls 晷刻之期, as opposed to 尋常之期.

28. **When some are seen advancing and some retreating, it is a lure.**

Capt. Calthrop is hardly right in translating: "An advance, followed by sudden retirement." It is rather a case of feigned confusion. As Tu Mu says: 僞爲雜亂不整之狀.

29. **When the soldiers stand leaning on their spears, they are faint from want of food.**

仗 is here probably not a synonym for 倚, but = 兵 "a weapon." The original text has 杖而立者, which has been corrected from the *T'ung Tien* and *Yü Lan*.

30. **If those who are sent to draw water begin by drinking themselves, the army is suffering from thirst.**

As Tu Mu remarks: 覩一人三軍可知也 "One may know the condition of a whole army from the behaviour of a single man." The 先 may mean either that they drink before drawing water for the army, or before they return to camp. Chang Yü takes the latter view. The *T'ung Tien* has the faulty reading 汲役先飲者, and the *Yü Lan*, worse still, 汲設飲者.

31. 見利而不進者勞也
32. 鳥集者虛也夜呼者恐也
33. 軍擾者將不重也旌旗動者亂也吏怒者倦也
34. 粟馬肉食軍無懸甀不返其舍者窮寇也

31. If the enemy sees an advantage to be gained

Not necessarily "booty," as Capt. Calthrop translates it. The *T'ung Tien* and *Yü Lan* read 向人見利, etc.

and makes no effort to secure it, the soldiers are exhausted.

32. If birds gather on any spot, it is unoccupied.

A useful fact to bear in mind when, for instance, as Ch'ên Hao says, the enemy has secretly abandoned his camp.

Clamour by night betokens nervousness.

Owing to false alarms; or, as Tu Mu explains it: 恐懼不安故夜呼以自壯也 "Fear makes men restless; so they fall to shouting at night in order to keep up their courage." The *T'ung Tien* inserts 喧 before 呼.

33. If there is disturbance in the camp, the general's authority is weak. If the banners and flags are shifted about, sedition is afoot.

The *T'ung Tien* and *Yü Lan* omit 旌.

If the officers are angry, it means that the men are weary.

And therefore, as Capt. Calthrop says, slow to obey. Tu Yu understands the sentence differently: "If all the officers of an army are angry with their general, it means that they are broken with fatigue" [owing to the exertions which he has demanded from them].

34. When an army feeds its horses with grain and kills its cattle for food,

粟馬肉食 is expanded by Mei Yao-ch'ên (following Tu Mu) into 給糧以秣乎馬殺畜以饗乎士, which is the sense I have given above. In the ordinary course of things, the men would be fed on grain and the horses chiefly on grass.

and when the men do not hang their cooking-pots

35. 諄諄翕翕徐言入入者失眾也

The *T'ung Tien* reads 缶, which is much the same as 皽, and the *Yü Lan* 笙, which is manifestly wrong.

over the camp-fires, showing that they will not return to their tents,

For 返, the *T'ung Tien* and *Yü Lan* both read 及.

you may know that they are determined to fight to the death.

For 窮寇, see VII. §36. I may quote here the illustrative passage from the *Hou Han Shu*, ch. 71, given in abbreviated form by the *P'ei Wên Yün Fu:* "The rebel 王國 Wang Kuo of 梁 Liang was besieging the town of 陳倉 Ch'ên-ts'ang, and 皇甫嵩 Huang-fu Sung, who was in supreme command, and 董卓 Tung Cho were sent out against him. The latter pressed for hasty measures, but Sung turned a deaf ear to his counsel. At last the rebels were utterly worn out, and began to throw down their weapons of their own accord. Sung was now for advancing to the attack, but Cho said: 'It is a principle of war not to pursue desperate men and not to press a retreating host.' Sung answered: 'That does not apply here. What I am about to attack is a jaded army, not a retreating host; with disciplined troops I am falling on a disorganised multitude, not a band of desperate men.' Thereupon he advanced to the attack unsupported by his colleague, and routed the enemy, Wang Kuo being slain." The inferior reading of the *T'u Shu* for § 34 is as follows: 殺馬肉食者軍無糧也懸皽不返其舍者窮寇也. The first clause strikes me as rather shallow for Sun Tzŭ, and it is hard to make anything of 懸皽 in the second without the negative. Capt. Calthrop, nothing daunted, set down in his first edition: "When they *cast away* their cooking-pots." He now has: "When the cooking-pots are hung up on the wall."

35. The sight of men whispering together

諄諄 is well explained by Tu Mu as 乏氣聲促 "speaking with bated breath."

in small knots

The *Shuo Wên* rather strangely defines 翕 by the word 起, but the *Êrh Ya* says 合 "to join" or "contract," which is undoubtedly its primary meaning. Chang Yü is right, then, in explaining it here by the word 聚. The other commentators are very much at sea: Ts'ao Kung says 失志貌, Tu Yu 不眞, Tu Mu 顚倒失次貌, Chia Lin 不安貌, Mei Yao-ch'ên 曠職事, Wang Hsi 患其上.

36. 屢賞者窘也數罰者困也
37. 先暴而後畏其衆者不精之至也
38. 來委謝者欲休息也

or speaking in subdued tones

入入 is said to be the same as 如如.

points to disaffection amongst the rank and file.

失衆 is equivalent to 失其衆心, the subject of course being "the general," understood. In the original text, which seems to be followed by several commentators, the whole passage stands thus: 諄諄翕翕徐與人言者失衆也. Here it would be the general who is talking to his men, not the men amongst themselves. For 翕, which is the chief stumbling-block in the way of this reading, the *T'u Shu* gives the very plausible emendation 謚 (also read *hsi*, and defined by K'ang Hsi as 疾言 "to speak fast"). But this is unnecessary if we keep to the standard text.

36. Too frequent rewards signify that the enemy is at the end of his resources;

Because, when an army is hard pressed, as Tu Mu says, there is always a fear of mutiny, and lavish rewards are given to keep the men in good temper.

too many punishments betray a condition of dire distress.

Because in such case discipline becomes relaxed, and unwonted severity is necessary to keep the men to their duty.

37. To begin by bluster, but afterwards to take fright at the enemy's numbers, shows a supreme lack of intelligence.

I follow the interpretation of Ts'ao Kung: 先輕敵後聞其衆則心惡之也, also adopted by Li Ch'üan, Tu Mu and Chang Yü. Another possible meaning, set forth by Tu Yu, Chia Lin, Mei Yao-ch'ên and Wang Hsi, is: "The general who is first tyrannical towards his men, and then in terror lest they should mutiny, etc." This would connect the sentence with what went before about rewards and punishments. The *T'ung Tien* and *Yü Lan* read 情 "affection" instead of 精.

38. When envoys are sent with compliments in their mouths, it is a sign that the enemy wishes for a truce.

39. 兵怒而相迎久而不合又不相去必謹察之
40. 兵非益多也惟無武進足以併力料敵取人
而已

Tu Mu says: 所以委質來謝此乃勢已窮或有他
故必欲休息也 "If the enemy open friendly relations by sending
hostages, it is a sign that they are anxious for an armistice, either because
their strength is exhausted or for some other reason." But it hardly
needs a Sun Tzŭ to draw such an obvious inference; and although Tu
Mu is supported by Mei Yao-ch'ên and Chang Yü, I cannot think that
hostages are indicated by the word 委.

39. If the enemy's troops march up angrily and remain
facing ours for a long time without either joining battle
or taking themselves off again, the situation is one that
demands great vigilance and circumspection.

Capt. Calthrop falls into a trap which often lurks in the word 相.
He translates: "When both sides, eager for a fight, face each other for a
considerable time, neither advancing nor retiring," etc. Had he reflected
a little, he would have seen that this is meaningless as addressed to a
commander who has control over the movements of his own troops.
相迎, then, does not mean that the two armies go to meet each other,
but simply that the other side comes up to us. Likewise with 相去.
If this were not perfectly clear of itself, Mei Yao-ch'ên's paraphrase would
make it so: 怒而來逆我, etc. As Ts'ao Kung points out, a
manœuvre of this sort may be only a *ruse* to gain time for an unexpected
flank attack or the laying of an ambush.

40. If our troops are no more in number than the
enemy, that is amply sufficient;

Wang Hsi's paraphrase, partly borrowed from Ts'ao Kung, is 權力
均足矣. Another reading, adopted by Chia Lin and the *T'u Shu*,
is 兵非貴益多, which Capt. Calthrop renders, much too loosely:
"Numbers are no certain mark of strength."

it only means that no direct attack can be made.

Literally, "no martial advance." That is to say, 正 "*chêng*" tactics
and frontal attacks must be eschewed, and stratagem resorted to instead.

What we can do is simply to concentrate all our available
strength, keep a close watch on the enemy, and obtain
reinforcements.

41. 夫惟無慮而易敵者必擒於人

42. 卒未親附而罰之則不服不服則難用也卒
已親附而罰不行則不可用也

This is an obscure sentence, and none of the commentators succeed in squeezing very good sense out of it. The difficulty lies chiefly in the words 取人, which have been taken in every possible way. I follow Li Ch'üan, who appears to offer the simplest explanation: 惟得人者勝也 "Only the side that gets more men will win." Ts'ao Kung's note, concise as usual to the verge of incomprehensibility, is 厮養足也. Fortunately we have Chang Yü to expound its meaning to us in language which is lucidity itself: 兵力既均又未見便雖未足剛進足以取人於厮養之中以并兵合力察敵而取勝不必假他兵以助己 "When the numbers are even, and no favourable opening presents itself, although we may not be strong enough to deliver a sustained attack, we can find additional recruits amongst our sutlers and camp-followers, and then, concentrating our forces and keeping a close watch on the enemy, contrive to snatch the victory. But we must avoid borrowing foreign soldiers to help us." He then quotes from Wei Liao Tzŭ, ch. 3: 助卒名為十萬其實不過數萬耳 "The nominal strength of mercenary troops may be 100,000, but their real value will be not more than half that figure." According to this interpretation, 取人 means "to get recruits," not from outside, but from the tag-rag and bobtail which follows in the wake of a large army. This does not sound a very soldierly suggestion, and I feel convinced that it is not what Sun Tzŭ meant. Chia Lin, on the other hand, takes the words in a different sense altogether, namely "to conquer the enemy" [cf. I. § 20]. But in that case they could hardly be followed by 而已. Better than this would be the rendering "to make isolated captures," as opposed to 武進 "a general attack."

41. He who exercises no forethought but makes light of his opponents is sure to be captured by them.

The force of 夫惟 is not easy to appreciate. Ch'ên Hao says 殊無遠慮但輕敵者, thus referring 惟 to the second verb. He continues, quoting from the Tso Chuan: 蜂蠆有毒而況國乎則小敵亦不可輕 "If bees and scorpions carry poison, how much more will a hostile state! [僖公, XXII. 3.] Even a puny opponent, then, should not be treated with contempt."

42. If soldiers are punished before they have grown

7

43. 故令之以文齊之以武是謂必取

44. 令素行以教其民則民服令不素行以教其
民則民不服

45. 令素信著者與眾相得也

attached to you, they will not prove submissive; and, unless submissive, they will be practically useless. If, when the soldiers have become attached to you, punishments are not enforced, they will still be useless.

This is wrongly translated by Capt. Calthrop: "If the troops know the general, but are not affected by his punishments, they are useless."

43. Therefore soldiers must be treated in the first instance with humanity, but kept under control by means of iron discipline.

文 and 武, according to Ts'ao Kung, are here equivalent to 仁 and 法 respectively. Compare our two uses of the word "civil." 晏子 Yen Tzŭ [† B.C. 493] said of 司馬穰苴 Ssŭ-ma Jang-chü: 文能附眾武能威敵也 "His civil virtues endeared him to the people; his martial prowess kept his enemies in awe." Cf. Wu Tzŭ, ch. 4 init.: 夫總文武者軍之將也兼剛柔者兵之事也 "The ideal commander unites culture with a warlike temper; the profession of arms requires a combination of hardness and tenderness." Again I must find fault with Capt. Calthrop's translation: "By humane treatment we obtain obedience; authority brings uniformity."

This is a certain road to victory.

44. If in training soldiers commands are habitually enforced, the army will be well-disciplined; if not, its discipline will be bad.

The T'ung Tien and Yü Lan read: 令素行以教其人者也令素行則人服令素不行則人不服.

45. If a general shows confidence in his men but always insists on his orders being obeyed,

The original text has 令素行者. 令素 is certainly awkward without 行, but on the other hand it is clear that Tu Mu accepted the T'ung Tien text, which is identical with ours. He says: "A general ought in time of peace to show kindly confidence in his men and also make his authority respected, so that when they come to face the enemy, orders may be executed and discipline maintained, because they all trust and look up to him." What Sun Tzŭ has said in § 44, however, would lead one rather to expect something like this: "If a general is always confident that his orders will be carried out," etc. Hence I am tempted to think that he may have written 令素信行者. But this is perhaps too conjectural.

the gain will be mutual.

Chang Yü says: 上以信使民民以信服上是上下相得也 "The general has confidence in the men under his command, and the men are docile, having confidence in him. Thus the gain is mutual." He quotes a pregnant sentence from Wei Liao Tzŭ, ch. 4: 令之之法小過無更小疑無中 "The art of giving orders is not to try to rectify minor blunders and not to be swayed by petty doubts." Vacillation and fussiness are the surest means of sapping the confidence of an army. Capt. Calthrop winds up the chapter with a final mistranslation of a more than usually heinous description: "Orders are always obeyed, if general and soldiers are in sympathy." Besides inventing the latter half of the sentence, he has managed to invert protasis and apodosis.

X. 地形篇

1. 孫子曰地形有通者有挂者有支者有隘者有險者有遠者

X. TERRAIN.

Only about a third of the chapter, comprising §§ 1—13, deals with 地形, the subject being more fully treated in ch. XI. The "six calamities" are discussed in §§ 14—20, and the rest of the chapter is again a mere string of desultory remarks, though not less interesting, perhaps, on that account.

1. Sun Tzŭ said: We may distinguish six kinds of terrain, to wit: (1) Accessible ground;

Mei Yao-ch'ên says: 道路交達 "plentifully provided with roads and means of communication."

(2) entangling ground;

The same commentator says: 網羅之地往必掛綴 "Net-like country, venturing into which you become entangled."

(3) temporising ground;

Tu Yu explains 支 as 久. This meaning is still retained in modern phrases such as 支拕, 支演 "stave off," "delay." I do not know why Capt. Calthrop calls 支地 "suspended ground," unless he is confusing it with 挂地.

(4) narrow passes; (5) precipitous heights;

The root idea in 隘 is narrowness; in 險, steepness.

(6) positions at a great distance from the enemy.

It is hardly necessary to point out the faultiness of this classification. A strange lack of logical perception is shown in the Chinaman's unquestioning acceptance of glaring cross-divisions such as the above.

2. 我可以往彼可以來日通

3. 通形者先居高陽利糧道以戰則利

2. Ground which can be freely traversed by both sides is called *accessible*.

Generally speaking, 平陸 "level country" is meant. Cf. IX. § 9: 處易.

3. With regard to ground of this nature,

The *T'ung Tien* reads 居通地.

be before the enemy in occupying the raised and sunny spots,

See IX. § 2. The *T'ung Tien* reads 先據其地.

and carefully guard your line of supplies.

A curious use of 利 as a verb, if our text is right. The general meaning is doubtless, as Tu Yu says, 無使敵絶己糧道 "not to allow the enemy to cut your communications." Tu Mu, who was not a soldier and can hardly have had any practical experience of fighting, goes more into detail and speaks of protecting the line of communications by a wall (壘), or enclosing it by embankments on each side (作甬道)! In view of Napoleon's dictum, "the secret of war lies in the communications,"* we could wish that Sun Tzŭ had done more than skirt the edge of this important subject here and in I. § 10, VII. § 11. Col. Henderson says: "The line of supply may be said to be as vital to the existence of an army as the heart to the life of a human being. Just as the duellist who finds his adversary's point menacing him with certain death, and his own guard astray, is compelled to conform to his adversary's movements, and to content himself with warding off his thrusts, so the commander whose communications are suddenly threatened finds himself in a false position, and he will be fortunate if he has not to change all his plans, to split up his force into more or less isolated detachments, and to fight with inferior numbers on ground which he has not had time to prepare, and where defeat will not be an ordinary failure, but will entail the ruin or the surrender of his whole army."**

Then you will be able to fight with advantage.

Omitted by Capt. Calthrop.

---

* See "Pensées de Napoléon Ier," no. 47.
** "The Science of War," chap. 2.

4, 可以往難以返曰挂

5. 挂形者敵無備出而勝之敵若有備出而不
勝難以返不利

6. 我出而不利彼出而不利曰支

7. 支形者敵雖利我我無出也引而去令敵半
出而擊之利

4. Ground which can be abandoned but is hard to re-occupy is called *entangling*.

Capt. Calthrop is wrong in translating 返 "retreat from it."

5. From a position of this sort, if the enemy is unprepared, you may sally forth and defeat him. But if the enemy is prepared for your coming, and you fail to defeat him, then, return being impossible, disaster will ensue.

不利 (an example of litotes) is paraphrased by Mei Yao-ch'ên as 必受制 "you will receive a check."

6. When the position is such that neither side will gain by making the first move, it is called *temporising* ground.

俱不便久相持也 "Each side finds it inconvenient to move, and the situation remains at a deadlock" (Tu Yu).

7. In a position of this sort, even though the enemy should offer us an attractive bait,

Tu Yu says 佯背我去 "turning their backs on us and pretending to flee." But this is only one of the lures which might induce us to quit our position. Here again 利 is used as a verb, but this time in a different sense: "to hold out an advantage to."

it will be advisable not to stir forth, but rather to retreat, thus enticing the enemy in his turn; then, when part of his army has come out, we may deliver our attack with advantage.

Mei Yao-ch'ên paraphrases the passage in a curious jingle, the scheme of rhymes being *abcbdd:* 各居所險、先出必敗、利而誘我、我不可愛、僞去引敵、半出而擊.

8. 隘形者我先居之必盈之以待敵
9. 若敵先居之盈而勿從不盈而從之
10. 險形者我先居之必居高陽以待敵

8. With regard to *narrow passes*, if you can occupy them first,

Capt. Calthrop says: "Defiles, make haste to occupy." But this is a conditional clause, answering to 若敵先居之 in the next paragraph.

let them be strongly garrisoned and await the advent of the enemy.

Because then, as Tu Yu observes, 皆制在我然後出奇以制敵 "the initiative will lie with us, and by making sudden and unexpected attacks we shall have the enemy at our mercy." The commentators make a great pother about the precise meaning of 盈, which to the foreign reader seems to present no difficulty whatever.

9. Should the enemy forestall you in occupying a pass, do not go after him if the pass is fully garrisoned, but only if it is weakly garrisoned.

10. With regard to *precipitous heights*, if you are beforehand with your adversary, you should occupy the raised and sunny spots, and there wait for him to come up.

Ts'ao Kung says: 地形險隘尤不可致於人 "The particular advantage of securing heights and defiles is that your actions cannot then be dictated by the enemy." [For the enunciation of the grand principle alluded to, see VI. § 2]. Chang Yü tells the following anecdote of 裴行儉 P'ei Hsing-chien (A.D. 619—682), who was sent on a punitive expedition against the Turkic tribes. "At nightfall he pitched his camp as usual, and it had already been completely fortified by wall and ditch, when suddenly he gave orders that the army should shift its quarters to a hill near by. This was highly displeasing to his officers, who protested loudly against the extra fatigue which it would entail on the men. P'ei Hsing-chien, however, paid no heed to their remonstrances and had the camp moved as quickly as possible. The same night, a terrific storm came on, which flooded their former place of encampment to the depth of over twelve feet. The recalcitrant officers were amazed at the sight, and owned that they had been in the wrong. 'How did you know what was going to happen?' they asked. P'ei Hsing-chien replied: 'From this time forward be content to obey orders without asking

11. 若敵先居之引而去之勿從也
12. 遠形者勢均難以挑戰戰而不利
13. 凡此六者地之道也將之至任不可不察也

unnecessary questions.' [See *Chiu T‘ang Shu*, ch. 84, fol. 12 *r*°., and *Hsin T‘ang Shu* ch. 108, fol. 5 *v*°.] From this it may be seen," Chang Yü continues, "that high and sunny places are advantageous not only for fighting, but also because they are immune from disastrous floods."

11. If the enemy has occupied them before you, do not follow him, but retreat and try to entice him away.

The turning-point of 李世民 Li Shih-min's campaign in 621 A.D. against the two rebels, 竇建德 Tou Chien-tê, King of 夏 Hsia, and 王世充 Wang Shih-ch‘ung, Prince of 鄭 Chêng, was his seizure of the heights of 武牢 Wu-lao, in spite of which Tou Chien-tê persisted in his attempt to relieve his ally in Lo-yang, was defeated and taken prisoner. [See *Chiu T‘ang Shu*, ch. 2, fol. 5 *v*°., and also ch. 54.]

12. If you are situated at a great distance from the enemy, and the strength of the two armies is equal,

The *T‘ung Tien* reads 夫通形均勢.

it is not easy to provoke a battle,

Ts‘ao Kung says that 挑戰 means 延敵 "challenging the enemy." But the enemy being far away, that plainly involves, as Tu Yu says, 迎敵 "going to meet him." The point of course is, that we must not think of undertaking a long and wearisome march, at the end of which 是我困敵銳 "we should be exhausted and our adversary fresh and keen."

and fighting will be to your disadvantage.

13. These six are the principles connected with Earth.

Or perhaps, "the principles relating to ground." See, however, I. § 8.

The general who has attained a responsible post must be careful to study them.

Capt. Calthrop omits 至任. Out of the foregoing six 地形, it will be noticed that nos. 3 and 6 have really no reference to the configuration of the country, and that only 4 and 5 can be said to convey any definite geographical idea.

14. 故兵有走者有弛者有陷者有崩者有亂者
有北者凡此六者非天之災將之過也
15. 夫勢均以一擊十曰走
16. 卒強吏弱曰弛吏強卒弱曰陷

14. Now an army is exposed to six several calamities, not arising from natural causes,

The *T'u Shu* reads 天地之災.

but from faults for which the general is responsible. These are: (1) Flight; (2) insubordination; (3) collapse; (4) ruin; (5) disorganisation; (6) rout.

I take exception to Capt. Calthrop's rendering of 陷 and 崩 as "distress" and "disorganisation," respectively.

15. Other conditions being equal, if one force is hurled against another ten times its size, the result will be the *flight* of the former.

Cf. III. § 10. The general's fault here is that of 不料力 "not calculating the enemy's strength." It is obvious that 勢 cannot have the same force as in § 12, where it was equivalent to 兵力. I should not be inclined, however, to limit it, with Chang Yü, to 將之智勇 兵之利銳 "the wisdom and valour of the general and the sharpness of the weapons." As Li Ch'üan very justly remarks, 若得形便 之地用奇伏之計則可矣 "Given a decided advantage in position, or the help of some stratagem such as a flank attack or an ambuscade, it would be quite possible [to fight in the ratio of one to ten]."

16. When the common soldiers are too strong and their officers too weak, the result is *insubordination*.

弛 "laxity" — the metaphor being taken from an unstrung bow. Capt. Calthrop's "relaxation" is not good, on account of its ambiguity. Tu Mu cites the unhappy case of 田布 T'ien Pu [*Hsin T'ang Shu*, ch. 148], who was sent to 魏 Wei in 821 A.D. with orders to lead an army against 王廷湊 Wang T'ing-ts'ou. But the whole time he was in command, his soldiers treated him with the utmost contempt, and openly flouted his authority by riding about the camp on donkeys, several thousands at a time. T'ien Pu was powerless to put a stop to this conduct, and when,

17. 大吏怒而不服遇敵懟而自戰將不知其能
日崩

after some months had passed, he made an attempt to engage the enemy,
his troops turned tail and dispersed in every direction. After that, the
unfortunate man committed suicide by cutting his throat.

**When the officers are too strong and the common soldiers
too weak, the result is** *collapse.*

Ts'ao Kung says: 吏強欲進卒弱輒陷 "The officers are
energetic and want to press on, the common soldiers are feeble and suddenly collapse." Note that 弱 is to be taken literally of physical weakness, whereas in the former clause it is figurative. Li Ch'üan makes 陷
equivalent to 敗, and Tu Mu explains it as 陷沒於死地 "stumb-
ling into a death-trap."

17. When the higher officers

大吏, according to Ts'ao Kung, are the 小將 "generals of inferior rank." But Li Ch'üan, Ch'ên Hao and Wang Hsi take the term
as simply convertible with 將 or 大將.

**are angry and insubordinate, and on meeting the enemy
give battle on their own account from a feeling of resentment, before the commander-in-chief can tell whether or
no he is in a position to fight, the result is** *ruin.*

Ts'ao Kung makes 大將, understood, the subject of 怒, which
seems rather far-fetched. Wang Hsi's note is: 謂將怒不以理
且不知裨佐之才激致其兇懟如山之崩壞也
"This means, the general is angry without just cause, and at the same
time does not appreciate the ability of his subordinate officers; thus he
arouses fierce resentment and brings an avalanche of ruin upon his head."
He takes 能, therefore, in the sense of 才; but I think that Ch'ên Hao
is right in his paraphrase 不顧能否 "they don't care if it be possible or no." My interpretation of the whole passage is that of Mei Yao-
ch'ên and Chang Yü. Tu Mu gives a long extract from the *Tso Chuan*,
宣公, XII. 3, showing how the great battle of 邲 Pi [597 B.C.] was
lost for the 晉 Chin State through the contumacy of 先縠 Hsien Hu
and the resentful spite of 魏錡 Wei I and 趙旃 Chao Chan. Chang
Yü also alludes to the mutinous conduct of 欒黶 Luan Yen [*ibid.*
襄公, XIV. 3].

18. 將弱不嚴教道不明吏卒無常陳兵縱橫曰
亂

19. 將不能料敵以少合衆以弱擊強兵無選鋒
曰北

18. When the general is weak and without authority;
when his orders are not clear and distinct;

Wei Liao Tzǔ (ch. 4) says: 上無疑令、則衆不二聽、
動無疑事、則衆不二志 "If the commander gives his orders
with decision, the soldiers will not wait to hear them twice; if his moves
are made without vacillation, the soldiers will not be in two minds about
doing their duty." General Baden-Powell says, italicising the words:
"The secret of getting successful work out of your trained men lies in
one nutshell — in the clearness of the instructions they receive." * Assuming that clear instructions beget confidence, this is very much what
Wei Liao Tzǔ (loc. cit.) goes on to say: 未有不信其心而能
得其力者也. Cf. also Wu Tzǔ ch. 3: 用兵之害猶豫
最大三軍之災生於狐疑 "the most fatal defect in a mili-
tary leader is diffidence; the worst calamities that befall an army arise
from hesitation."

when there are no fixed duties assigned to officers and men,

吏卒皆不拘常度 "Neither officers nor men have any regular
routine" [Tu Mu].

and the ranks are formed in a slovenly haphazard manner,
the result is utter *disorganisation.*

19. When a general, unable to estimate the enemy's
strength, allows an inferior force to engage a larger one,
or hurls a weak detachment against a powerful one, and
neglects to place picked soldiers in the front rank, the
result must be a *rout.*

Chang Yü paraphrases the latter part of the sentence 不選驍勇
之士使爲先鋒兵必敗北也, and continues: 凡戰必
用精銳爲前鋒者一則壯吾志一則挫敵威也
"Whenever there is fighting to be done, the keenest spirits should be

* "Aids to Scouting," p. xii.

20. 凡此六者敗之道也將之至任不可不察也
21. 夫地形者兵之助也料敵制勝計險阸遠近
上將之道也

appointed to serve in the front ranks, both in order to strengthen the resolution of our own men and to demoralise the enemy." Cf. the *primi ordines* of Caesar ("De Bello Gallico," V. 28, 44 *et al.*). There seems little to distinguish 北 from 走 in § 15, except that 北 is a more forcible word.

20. **These are six ways of courting defeat,**

Ch'ên Hao makes them out to be: (1) 不量寡眾 "neglect to estimate the enemy's strength;" (2) 本乏刑德 "want of authority;" (3) 失於訓練 "defective training;" (4) 非理興怒 "unjustifiable anger;" (5) 法令不行 "non-observance of discipline;" (6) 不擇驍果 "failure to use picked men."

which must be carefully noted by the general who has attained a responsible post.

See *supra*, § 13.

21. **The natural formation of the country is the soldier's best ally;**

Chia Lin's text has the reading 易 for 助. Ch'ên Hao says: 天時不如地利 "The advantages of weather and season are not equal to those connected with ground."

but a power of estimating the adversary,

The insertion of a "but" is necessary to show the connection of thought here. A general should always utilise, but never rely wholly on natural advantages of terrain.

of controlling the forces of victory,

制勝 is one of those condensed expressions which mean so much in Chinese, and so little in an English translation. What it seems to imply is complete mastery of the situation from the beginning.

and of shrewdly calculating difficulties, dangers and distances,

The *T'ung Tien* and *Yü Lan* read 計極險易利害遠近. I am decidedly puzzled by Capt. Calthrop's translation: "an eye for steepness, *command* and distances." Where did he find the word which I have put in italics?

22. 知此而用戰者必勝不知此而用戰者必敗
23. 故戰道必勝主曰無戰必戰可也戰道不勝
主曰必戰無戰可也

constitutes the test of a great general.

A somewhat free translation of 道. As Chang Yü remarks, these are 兵之本 "the essentials of soldiering," ground being only a helpful accessory.

22. He who knows these things, and in fighting puts his knowledge into practice, will win his battles. He who knows them not, nor practises them, will surely be defeated.

23. If fighting is sure to result in victory, then you must fight, even though the ruler forbid it; if fighting will not result in victory, then you must not fight even at the ruler's bidding.

Cf. VIII. § 3 *fin.* Huang Shih-kung of the Ch'in dynasty, who is said to have been the patron of 張良 Chang Liang and to have written the 三略, has these words attributed to him: 出軍行師將在 自專進退內御則功難成故聖主明王跪而推 轂 "The responsibility of setting an army in motion must devolve on the general alone; if advance and retreat are controlled from the Palace, brilliant results will hardly be achieved. Hence the god-like ruler and the enlightened monarch are content to play a humble part in furthering their country's cause [*lit.*, kneel down to push the chariot wheel]." This means that 閫外之事將軍裁之 "in matters lying outside the zenana, the decision of the military commander must be absolute." Chang Yü also quotes the saying: 軍中不聞天子之詔 "Decrees of the Son of Heaven do not penetrate the walls of a camp." Napoleon, who has been accused of allowing his generals too little independence of action, speaks in the same sense: "Un général en chef n'est pas à couvert de ses fautes à la guerre par un ordre de son souverain ou du ministre, quand celui qui le donne est éloigné du champ d'opération, et qu'il connaît mal, ou ne connaît pas du tout le dernier état des choses." *

---

* "Maximes de Guerre," no. 72.

24. 故進不求名退不避罪唯民是保而利合於
主國之寶也

25. 視卒如嬰兒故可與之赴深谿視卒如愛子
故可與之俱死

24. The general who advances without coveting fame and retreats without fearing disgrace,

It was Wellington, I think, who said that the hardest thing of all for a soldier is to retreat.

whose only thought is to protect his country and do good service for his sovereign,

合, which is omitted by the *T'u Shu*, is said by Ch'ên Hao to be equivalent to 歸. If it had to be separately translated, it would be something like our word "accrue."

is the jewel of the kingdom.

A noble presentment, in few words, of the Chinese "happy warrior." Such a man, says Ho Shih, 罪及其身不悔也 "even if he had to suffer punishment, would not regret his conduct."

25. Regard your soldiers as your children, and they will follow you into the deepest valleys; look on them as your own beloved sons, and they will stand by you even unto death.

Cf. I. § 6. In this connection, Tu Mu draws for us an engaging picture of the famous general Wu Ch'i, from whose treatise on war I have frequently had occasion to quote: "He wore the same clothes and ate the same food as the meanest of his soldiers, refused to have either a horse to ride or a mat to sleep on, carried his own surplus rations wrapped in a parcel, and shared every hardship with his men. One of his soldiers was suffering from an abscess, and Wu Ch'i himself sucked out the virus. The soldier's mother, hearing this, began wailing and lamenting. Somebody asked her, saying: 'Why do you cry? Your son is only a common soldier, and yet the commander-in-chief himself has sucked the poison from his sore.' The woman replied: 'Many years ago, Lord Wu performed a similar service for my husband, who never left him afterwards, and finally met his death at the hands of the enemy. And now that he has done the same for my son, he too will fall fighting I know not where'." Li Ch'üan mentions 楚子 the Viscount of Ch'u, who invaded the small state of 蕭 Hsiao during the winter. 申公 The Duke of Shên said to him: "Many of

26. 厚而不能使愛而不能令亂而不能治譬如
驕子不可用也

27. 知吾卒之可以擊而不知敵之不可擊勝之
半也

the soldiers are suffering severely from the cold." So he made a round of the whole army, comforting and encouraging the men; and straightway they felt as if they were clothed in garments lined with floss silk. [*Tso Chuan*, 宣公, XII. 5]. Chang Yü alludes to the same passage, saying: 溫言一撫士同挾纊.

26. If, however, you are indulgent, but unable to make your authority felt; kind-hearted, but unable to enforce your commands; and incapable, moreover, of quelling disorder:

Capt. Calthrop has got these three clauses quite wrong. The last he translates: "overindulgence may produce disorder."

then your soldiers must be likened to spoilt children; they are useless for any practical purpose.

Cf. IX. § 42. We read in the 陰符經, pt. 2: 害生于恩 "Injury comes out of kindness." Li Ching once said that if you could make your soldiers afraid of you, they would not be afraid of the enemy. Tu Mu recalls an instance of stern military discipline which occurred in 219 A.D., when 呂蒙 Lü Mêng was occupying the town of 江陵 Chiang-ling. He had given stringent orders to his army not to molest the inhabitants nor take anything from them by force. Nevertheless, a certain officer serving under his banner, who happened to be a fellow-townsman, ventured to appropriate a bamboo hat (笠) belonging to one of the people, in order to wear it over his regulation helmet as a protection against the rain. Lü Mêng considered that the fact of his being also a native of 汝南 Ju-nan should not be allowed to palliate a clear breach of discipline, and accordingly he ordered his summary execution, the tears rolling down his face, however, as he did so. This act of severity filled the army with wholesome awe, and from that time forth even articles dropped in the highway were not picked up. [*San Kuo Chih*, ch. 54, f. 13 r°. & v°.].

27. If we know that our own men are in a condition to attack, but are unaware that the enemy is not open to attack, we have gone only halfway towards victory.

That is, as Ts'ao Kung says, "the issue in this case is uncertain."

28. 知敵之可擊而不知吾卒之不可以擊勝之半也

29. 知敵之可擊知吾卒之可以擊而不知地形之不可以戰勝之半也

30. 故知兵者動而不迷舉而不窮

31. 故曰知彼知己勝乃不殆知地知天勝乃可全

28. If we know that the enemy is open to attack, but are unaware that our own men are not in a condition to attack, we have gone only halfway towards victory.

Cf. III. § 13 (1).

29. If we know that the enemy is open to attack, and also know that our men are in a condition to attack, but are unaware that the nature of the ground makes fighting impracticable, we have still gone only halfway towards victory.

I may take this opportunity of pointing out the rather nice distinction in meaning between 擊 and 攻. The latter is simply "to attack" without any further implication, whereas 擊 is a stronger word which in nine cases out of ten means "to attack with expectation of victory," "to fall upon," as we should say, or even "to crush." On the other hand, 擊 is not quite synonymous with 伐, which is mostly used of operations on a larger scale, as of one State *making war* on another, often with the added idea of invasion. 征, finally, has special reference to the subjugation of rebels. See Mencius, VII. 2. ii. 2.

30. Hence the experienced soldier, once in motion, is never bewildered; once he has broken camp, he is never at a loss.

The reason being, according to Tu Mu, that he has taken his measures so thoroughly as to ensure victory beforehand. "He does not move recklessly," says Chang Yü, "so that when he does move, he makes no mistakes." Another reading substitutes 困 for 迷 and 頓 for 窮. The latter variant only is adopted by the *T'ung Tien* and *Yü Lan*. Note that 窮 here means "at the end of his *mental* resources."

31. Hence the saying: If you know the enemy and know yourself, your victory will not stand in doubt;

Capt. Calthrop makes the saying end here, which cannot be justified.

if you know Heaven and know Earth,

天 and 地 are transposed for the sake of the jingle between 天 and 全. The original text, however, has 知 天 知 地, and the correction has been made from the *T'ung Tien*.

you may make your victory complete.

As opposed to 勝 之 半, above. The original text has 勝 乃 不 窮, the corruption being perhaps due to the occurrence of 不 窮 in the preceding sentence. Here, however 不 窮 would not be synonymous with 不 困, but equivalent to 不 可 以 窮 "inexhaustible," "beyond computation." Cf. V. § 11. The *T'ung Tien* has again supplied the true reading. Li Ch'üan sums up as follows: 人 事 天 時 地 利 三 者 同 知 則 百 戰 百 勝 "Given a knowledge of three things — the affairs of man, the seasons of heaven and the natural advantages of earth —, victory will invariably crown your battles."

XI. 九地篇

1. 孫子曰用兵之法有散地有輕地有爭地有交地有衢地有重地有圮地有圍地有死地
2. 諸侯自戰其地者爲散地

XI. THE NINE SITUATIONS.

Li Ch'üan is not quite right in calling these 勝敵之地. As we shall see, some of them are highly disadvantageous from the military point of view. Wang Hsi more correctly says: 用兵之地利害有九也 "There are nine military situations, good and bad." One would like to distinguish the 九地 from the six 地形 of chap. X by saying that the latter refer to the natural formation or geographical features of the country, while the 九地 have more to do with the condition of the army, being 地勢 "situations" as opposed to "grounds." But it is soon found impossible to carry out the distinction. Both are cross-divisions, for among the 地形 we have "temporising ground" side by side with "narrow passes," while in the present chapter there is even greater confusion.

1. Sun Tzǔ said: The art of war recognises nine varieties of ground: (1) Dispersive ground; (2) facile ground; (3) contentious ground; (4) open ground; (5) ground of intersecting highways; (6) serious ground; (7) difficult ground; (8) hemmed-in ground; (9) desperate ground.

2. When a chieftain is fighting in his own territory, it is dispersive ground.

So called because the soldiers, being near to their homes and auxious to see their wives and children, are likely to seize the opportunity afforded by a battle and scatter in every direction. "In their advance," observes Tu Mu, "they will lack the valour of desperation, and when they retreat, they will find harbours of refuge." The 者, which appears in the *T'u Shu*, seems to have been accidentally omitted in my edition of the standard text.

3. 入人之地而不深者爲輕地
4. 我得則利彼得亦利者爲爭地

3. When he has penetrated into hostile territory, but to no great distance, it is facile ground.

Li Ch'üan and Ho Shih say 輕於退也 "because of the facility for retreating," and the other commentators give similar explanations. Tu Mu remarks: 師出越境必焚舟梁示民無返顧之心 "When your army has crossed the border, you should burn your boats and bridges, in order to make it clear to everybody that you have no hankering after home." I do not think that "disturbing ground," Capt. Calthrop's rendering of 輕地, has anything to justify it. If an idiomatic translation is out of the question, one should at least attempt to be literal.

4. Ground the possession of which imports great advantage to either side, is contentious ground.

I must apologise for using this word in a sense not known to the dictionary, i.e. "to be contended for" — Tu Mu's 必爭之地. Ts'ao Kung says: 可以少勝衆弱勝强 "ground on which the few and the weak can defeat the many and the strong," such as 阨喉 "the neck of a pass," instanced by Li Ch'üan. Thus, Thermopylae was a 爭地, because the possession of it, even for a few days only, meant holding the entire invading army in check and thus gaining invaluable time. Cf. Wu Tzŭ, ch. V. *ad init.*: 以一擊十莫善於阨 "For those who have to fight in the ratio of one to ten, there is nothing better than a narrow pass." When 呂光 Lü Kuang was returning from his triumphant expedition to Turkestan in 385 A.D., and had got as far as 宜禾 I-ho, laden with spoils, 梁熙 Liang Hsi, administrator of 涼州 Liang-chou, taking advantage of the death of Fu Chien, King of Ch'in, plotted against him and was for barring his way into the province. 楊翰 Yang Han, governor of 高昌 Kao-ch'ang, counselled him, saying: "Lü Kuang is fresh from his victories in the west, and his soldiers are vigorous and mettlesome. If we oppose him in the shifting sands of the desert, we shall be no match for him, and we must therefore try a different plan. Let us hasten to occupy the defile at the mouth of the 高梧 Kao-wu pass, thus cutting him off from supplies of water, and when his troops are prostrated with thirst, we can dictate our own terms without moving. Or if you think that the pass I mention is too far off, we could make a stand against him at the 伊吾 I-wu pass, which is

116

5. 我可以往彼可以來者爲交地
6. 諸侯之地三屬先至而得天下之衆者爲衢
地

nearer. The cunning and resource of 子房 Tzǔ-fang himself [i.e. 張良] would be expended in vain against the enormous strength of these two positions." Liang Hsi, refusing to act on this advice, was overwhelmed and swept away by the invader. [See 晉書, ch. 122, fol. 3 rº, and 歷代紀事年表, ch. 43, fol. 26.]

5. Ground on which each side has liberty of movement is open ground.

This is only a makeshift translation of 交, which according to Ts'ao Kung stands for 交錯 "ground covered with a network of roads," like a chess-board. Another interpretation, suggested by Ho Shih, is 交通 "ground on which intercommunication is easy." In either case, it must evidently be 平原 "flat country," and therefore 不可杜絕 "cannot be blocked." Cf. 通形, X. § 2.

6. Ground which forms the key to three contiguous states,

我與敵相當而旁有他國也 "Our country adjoining the enemy's and a third country conterminous with both." [Ts'ao Kung.] Mêng Shih instances the small principality of 鄭 Chêng, which was bounded on the north-east by 齊 Ch'i, on the west by 晉 Chin, and on the south by 楚 Ch'u.

so that he who occupies it first has most of the Empire at his command,

天下 of course stands for the loose confederacy of states into which China was divided under the Chou dynasty. The belligerent who holds this dominating position can constrain most of them to become his allies. See infra, § 48. 衆 appears at first sight to be "the masses" or "population" of the Empire, but it is more probably, as Tu Yu says, 諸侯之衆.

is ground of intersecting highways.

Capt. Calthrop's "path-ridden ground" might stand well enough for 交地 above, but it does not bring out the force of 衢地, which clearly denotes the central position where important highways meet.

7. 入人之地深背城邑多者爲重地

8. 山林險阻沮澤凡難行之道者爲圯地

9. 所由入者隘所從歸者迂彼寡可以擊吾之
衆者爲圍地

10. 疾戰則存不疾戰則亡者爲死地

7. When an army has penetrated into the heart of a hostile country, leaving a number of fortified cities in its rear,

After 多, the *T'ung Tien* intercalates the gloss 難以返.

it is serious ground.

Wang Hsi explains the name by saying that 兵至此者事勢重也 "when an army has reached such a point, its situation is serious." Li Ch'üan instances (1) the victorious march of 樂毅 Yo I into the capital of Ch'i in 284 B.C., and (2) the attack on Ch'u, six years later, by the Ch'in general 白起 Po Ch'i.

8. Mountain forests,

Or simply, "forests." I follow the *T'u Shu* in omitting the 行 before 山林, given in the standard text, which is not only otiose but spoils the rhythm of the sentence.

rugged steeps, marshes and fens — all country that is hard to traverse: this is difficult ground.

圯 *p'i*[3] (to be distinguished from 圮 *i*[4]) is defined by K'ang Hsi (after the *Shuo Wên*) as 毀 "to destroy." Hence Chia Lin explains 圯地 as ground 經水所毀 "that has been ruined by water passing over it," and Tu Yu simply as 沮洳之地 "swampy ground." But Ch'ên Hao says that the word is specially applied to deep hollows — what Chu-ko Liang, he tells us, used to designate by the expressive term 地獄 "earth-hells." Compare the 天井 of IX. § 15.

9. Ground which is reached through narrow gorges, and from which we can only retire by tortuous paths, so that a small number of the enemy would suffice to crush a large body of our men: this is hemmed-in ground.

10. Ground on which we can only be saved from destruction by fighting without delay, is desperate ground.

11. 是 故 散 地 則 無 以 戰 輕 地 則 無 止 爭 地 則 無 攻

The situation, as pictured by Ts'ao Kung, is very similar to the 圍地, except that here escape is no longer possible: 前 有 高 山 後 有 大 水 進 則 不 得 退 則 有 礙 "A lofty mountain in front, a large river behind, advance impossible, retreat blocked." Ch'ên Hao says: 人 在 死 地 如 坐 漏 船 伏 燒 屋 "to be on 'desperate ground' is like sitting in a leaking boat or crouching in a burning house." Tu Mu quotes from Li Ching a vivid description of the plight of an army thus entrapped: "Suppose an army invading hostile territory without the aid of local guides: — it falls into a fatal snare and is at the enemy's mercy. A ravine on the left, a mountain on the right, a pathway so perilous that the horses have to be roped together and the chariots carried in slings, no passage open in front, retreat cut off behind, no choice but to proceed in single file (鴈 行 魚 貫 之 嚴). Then, before there is time to range our soldiers in order of battle, the enemy in overwhelming strength suddenly appears on the scene. Advancing, we can nowhere take a breathing-space; retreating, we have no haven of refuge. We seek a pitched battle, but in vain; yet standing on the defensive, none of us has a moment's respite. If we simply maintain our ground, whole days and months will crawl by; the moment we make a move, we have to sustain the enemy's attacks on front and rear. The country is wild, destitute of water and plants; the army is lacking in the necessaries of life, the horses are jaded and the men worn-out, all the resources of strength and skill unavailing, the pass so narrow that a single man defending it can check the onset of ten thousand; all means of offence in the hands of the enemy, all points of vantage already forfeited by ourselves: — in this terrible plight, even though we had the most valiant soldiers and the keenest of weapons, how could they be employed with the slightest effect?" Students of Greek history may be reminded of the awful close to the Sicilian expedition, and the agony of the Athenians under Nicias and Demosthenes. [*See* Thucydides, VII. 78 sqq.].

11. On dispersive ground, therefore, fight not. On facile ground, halt not. On contentious ground, attack not.

But rather let all your energies be bent on occupying the advantageous position first. So Ts'ao Kung. Li Ch'üan and others, however, suppose the meaning to be that the enemy has already forestalled us, so that it would be sheer madness to attack. In the 孫 子 敍 錄, when the King of Wu inquires what should be done in this case, Sun Tzŭ replies: "The rule with regard to contentious ground is that those in possession have the advantage over the other side. If a position of this kind is

12. 衢地則無絕衢地則合交
13. 重地則掠圯地則行

secured first by the enemy, beware of attacking him. Lure him away by pretending to flee — show your banners and sound your drums — make a dash for other places that he cannot afford to lose — trail brushwood and raise a dust — confound his ears and eyes — detach a body of your best troops, and place it secretly in ambuscade. Then your opponent will sally forth to the rescue."

12. On open ground, do not try to block the enemy's way.

Because the attempt would be futile, and would expose the blocking force itself to serious risks. There are two interpretations of 無絶. I follow that of Chang Yü (不可以兵阻絶其路). The other is indicated in Ts'ao Kung's brief note: 相及屬也 "Draw closer together" — i.e., see that a portion of your own army is not cut off. Wang Hsi points out that 衢地 is only another name for the 通地 "accessible ground" of X. § 2, and says that the advice here given is simply a variation of 利糧道 "keep a sharp eye on the line of supplies," be careful that your communications are not cut. The *T'ung Tien* reads 無相絶.

On ground of intersecting highways, join hands with your allies.

Or perhaps, "form alliances with neighbouring states." Thus Ts'ao Kung has: 結諸侯也. Capt. Calthrop's "cultivate intercourse" is much too timid and vague. The original text reads 衢合.

13. On serious ground, gather in plunder.

On this, Li Ch'üan has the following delicious note: 深入敵境 不可非義失人心如漢高祖入秦無犯婦女無 取寶貨得人心也此筌以掠字爲無掠字 "When an army penetrates far into the enemy's country, care must be taken not to alienate the people by unjust treatment. Follow the example of the Han Emperor Kao Tsu, whose march into Ch'in territory was marked by no violation of women or looting of valuables. [*Nota bene:* this was in 207 B.C., and may well cause us to blush for the Christian armies that entered Peking in 1900 A.D.] Thus he won the hearts of all. In the present passage, then, I think that the true reading must be, not 掠 'plunder,' but 無掠 'do not plunder'." Alas, I fear that in this instance the worthy commentator's feelings outran his judgment. Tu Mu,

14. 圍地則謀死地則戰
15. 所謂古之善用兵者能使敵人前後不相及
衆寡不相恃貴賤不相救上下不相扶

at least, has no such illusions. He says: "When encamped on 'serious ground,' there being no inducement as yet to advance further, and no possibility of retreat, one ought to take measures for a protracted resistance by bringing in provisions from all sides, and keep a close watch on the enemy." Cf. also II. §9: 因糧於敵.

In difficult ground, keep steadily on the march.

Or, in the words of VIII. §2, 無舍 "do not encamp."

14. On hemmed-in ground, resort to stratagem.

Ts'ao Kung says: 發奇謀 "Try the effect of some unusual artifice;" and Tu Yu amplifies this by saying: 居此則當權謀詭譎可以免難 "In such a position, some scheme must be devised which will suit the circumstances, and if we can succeed in deluding the enemy, the peril may be escaped." This is exactly what happened on the famous occasion when Hannibal was hemmed in among the mountains on the road to Casilinum, and to all appearances entrapped by the Dictator Fabius. The stratagem which Hannibal devised to baffle his foes was remarkably like that which T'ien Tan had also employed with success exactly 62 years before. [See IX. § 24, note.] When night came on, bundles of twigs were fastened to the horns of some 2000 oxen and set on fire, the terrified animals being then quickly driven along the mountain side towards the passes which were beset by the enemy. The strange spectacle of these rapidly moving lights so alarmed and discomfited the Romans that they withdrew from their position, and Hannibal's army passed safely through the defile. [See Polybius, III. 93, 94; Livy, XXII. 16, 17.]

On desperate ground, fight.

For, as Chia Lin remarks: 力戰或生守隅則死 "if you fight with all your might, there is a chance of life; whereas death is certain if you cling to your corner."

15. Those who were called skilful leaders of old

所謂 is omitted in the *T'u Shu* text.

knew how to drive a wedge between the enemy's front and rear;

More literally, "cause the front and rear to lose touch with each other."

16. 卒離而不集兵合而不齊
17. 合於利而動不合於利而止
18. 敢問敵眾整而將來待之若何曰先奪其所
愛則聽矣

to prevent co-operation between his large and small divisions; to hinder the good troops from rescuing the bad,

I doubt if 貴賤 can mean "officers and men," as Capt. Calthrop translates. This is wanted for 上 下.

the officers from rallying their men.

The reading 扶, derived from the *Yü Lan*, must be considered very doubtful. The original text has 救, and the *T'u Shu* 收.

16. When the enemy's men were scattered, they prevented them from concentrating;

Capt. Calthrop translates 卒離 "they scattered the enemy," which cannot be right.

even when their forces were united, they managed to keep them in disorder.

Mei Yao-ch'ên's note makes the sense plain: 或已離而不能合或雖合而不能齊. All these clauses, of course, down to 不齊, are dependent on 使 in § 15.

17. When it was to their advantage, they made a forward move; when otherwise, they stopped still.

Mei Yao-ch'ên connects this with the foregoing: 然能使敵若此當須有利則動無利則止 "Having succeeded in thus dislocating the enemy, they would push forward in order to secure any advantage to be gained; if there was no advantage to be gained, they would remain where they were."

18. If asked how to cope with a great host of the enemy in orderly array and on the point of marching to the attack,

敢問 is like 或問, introducing a supposed question.

I should say: "Begin by seizing something which your opponent holds dear; then he will be amenable to your will."

19. 兵之情主速乘人之不及由不虞之道攻其
所不戒也

Opinions differ as to what Sun Tzŭ had in mind. Ts'ao Kung thinks
it is 其所恃之利 "some strategical advantage on which the enemy
is depending." Tu Mu says: 據我便地畧我田野利其
糧道斯三者敵人之所愛惜倚恃者也 "The three
things which an enemy is anxious to do, and on the accomplishment of
which his success depends, are: (1) to capture our favourable positions;
(2) to ravage our cultivated land; (3) to guard his own communications."
Our object then must be to thwart his plans in these three directions
and thus render him helpless. [Cf. III. § 3.] But this exegesis unduly
strains the meaning of 奪 and 愛, and I agree with Ch'ên Hao, who
says that 所愛 does not refer only to strategical advantages, but is
any person or thing that may happen to be of importance to the enemy.
By boldly seizing the initiative in this way, you at once throw the other
side on the defensive.

19. Rapidity is the essence of war:

兵之情 means "the conditions of war," not, as Capt. Calthrop
says, "the spirit of the troops." According to Tu Mu, 此統言兵
之情狀 "this is a summary of leading principles in warfare," and
he adds: 此乃兵之深情將之至事也 "These are the
profoundest truths of military science, and the chief business of the general."
The following anecdotes, told by Ho Shih, show the importance attached
to speed by two of China's greatest generals. In 227 A.D., 孟達
Mêng Ta, governor of 新城 Hsin-ch'êng under the Wei Emperor Wên
Ti, was meditating defection to the House of Shu, and had entered into
correspondence with Chu-ko Liang, Prime Minister of that State. The
Wei general Ssŭ-ma I was then military governor of 宛 Wan, and getting wind of Mêng Ta's treachery, he at once set off with an army to
anticipate his revolt, having previously cajoled him by a specious message
of friendly import. Ssŭ-ma's officers came to him and said: "If Mêng
Ta has leagued himself with Wu and Shu, the matter should be thoroughly investigated before we make a move." Ssŭ-ma I replied: "Mêng
Ta is an unprincipled man, and we ought to go and punish him at once,
while he is still wavering and before he has thrown off the mask." Then, by
a series of forced marches, he brought his army under the walls of Hsin-
ch'êng within the space of eight days. Now Mêng Ta had previously said in
a letter to Chu-ko Liang: "Wan is 1200 li from here. When the news
of my revolt reaches Ssŭ-ma I, he will at once inform his Imperial Master,
but it will be a whole month before any steps can be taken, and by that

20. 凡為客之道深入則專主人不克
21. 掠於饒野三軍足食
22. 謹養而勿勞併氣積力運兵計謀為不可測

time my city will be well fortified. Besides, Ssŭ-ma I is sure not to come himself, and the generals that will be sent against us are not worth troubling about." The next letter, however, was filled with consternation: "Though only eight days have passed since I threw off my allegiance, an army is already at the city-gates. What miraculous rapidity is this!" A fortnight later, Hsin-ch'êng had fallen and Mêng Ta had lost his head. [See *Chin Shu*, ch. 1, f. 3.] In 621 A.D., Li Ching was sent from 夔州 K'uei-chou in Ssŭ-ch'uan to reduce the successful rebel 蕭銑 Hsiao Hsien, who had set up as Emperor at the modern 荆州 Ching-chou Fu in Hupeh. It was autumn, and the Yangtsze being then in flood, Hsiao Hsien never dreamt that his adversary would venture to come down through the gorges, and consequently made no preparations. But Li Ching embarked his army without loss of time, and was just about to start when the other generals implored him to postpone his departure until the river was in a less dangerous state for navigation. Li Ching replied: "To the soldier, overwhelming speed is of paramount importance, and he must never miss opportunities. Now is the time to strike, before Hsiao Hsien even knows that we have got an army together. If we seize the present moment when the river is in flood, we shall appear before his capital with startling suddenness, like the thunder which is heard before you have time to stop your ears against it. [*See* VII, § 19, note.] This is the great principle in war. Even if he gets to know of our approach, he will have to levy his soldiers in such a hurry that they will not be fit to oppose us. Thus the full fruits of victory will be ours." All came about as he predicted, and Hsiao Hsien was obliged to surrender, nobly stipulating that his people should be spared and he alone suffer the penalty of death. [See *Hsin T'ang Shu*, ch. 93, f. 1 *v°*.]

take advantage of the enemy's unreadiness, make your way by unexpected routes, and attack unguarded spots.

20. The following are the principles to be observed by an invading force: The further you penetrate into a country, the greater will be the solidarity of your troops, and thus the defenders will not prevail against you.

21. Make forays in fertile country in order to supply your army with food.

Cf. *supra*, § 13. Li Ch'üan does not venture on a note here.

22. Carefully study the well-being of your men,

謹養, according to Wang Hsi, means: 撫循飲食周謹之 "Pet them, humour them, give them plenty of food and drink, and look after them generally."

and do not overtax them. Concentrate your energy and hoard your strength.

Tu Mu explains these words in a rhyming couplet: 氣全力盛 一發取勝; and Ch'ên recalls the line of action adopted in 224 B.C. by the famous general 王翦 Wang Chien, whose military genius largely contributed to the success of the First Emperor. He had invaded the Ch'u State, where a universal levy was made to oppose him. But, being doubtful of the temper of his troops, he declined all invitations to fight and remained strictly on the defensive. In vain did the Ch'u general try to force a battle: day after day Wang Chien kept inside his walls and would not come out, but devoted his whole time and energy to winning the affection and confidence of his men. He took care that they should be well fed, sharing his own meals with them, provided facilities for bathing, and employed every method of judicious indulgence to weld them into a loyal and homogeneous body. After some time had elapsed, he told off certain persons to find out how the men were amusing them-selves. The answer was, that they were contending with one another in putting the weight and long-jumping (投石超距). When Wang Chien heard that they were engaged in these athletic pursuits, he knew that their spirits had been strung up to the required pitch and that they were now ready for fighting. By this time the Ch'u army, after repeating their challenge again and again, had marched away eastwards in disgust. The Ch'in general immediately broke up his camp and followed them, and in the battle that ensued they were routed with great slaughter. Shortly afterwards, the whole of Ch'u was conquered by Ch'in, and the king 負芻 Fu-ch'u led into captivity. [See *Shih Chi*, ch. 73, f. 5 r°. It should be noted that, 楚 being a taboo character under the Ch'in dynasty, the name figures as 荆 throughout.]

Keep your army continually on the move,

In order that the enemy may never know exactly where you are. It has struck me, however, that the true reading might be, not 運兵, but 連兵 "link your army together" [cf. *supra* § 46, 吾將使之屬], which would be more in keeping with 併氣積力. Capt. Calthrop cuts the Gordian knot by omitting the words altogether.

and devise unfathomable plans.

Ch'ang Yü's paraphrase is: 常爲不可測度之計.

23. 投之無所往死且不北死焉不得士人盡力
24. 兵士甚陷則不懼無所往則固深入則拘不
　　得已則鬥

23. Throw your soldiers into positions whence there is no escape, and they will prefer death to flight.

Cf. Nicias' speech to the Athenians: Τό τε ξύμπαν γνῶτε, ὦ ἄνδρες στρατιῶται, ἀναγκαῖόν τε ὃν ὑμῖν ἀνδράσιν ἀγαθοῖς γίγνεσθαι, ὡς μὴ ὄντος χωρίου ἐγγὺς ὅποι ἂν μαλακισθέντες σωθεῖτε, etc. [Thuc. VII. 77. vii.]

If they will face death, there is nothing they may not achieve.

死 by itself constitutes the protasis, and 焉 is the interrogative = 安. Capt. Calthrop makes the protasis end with 得: "If there be no alternative but death." But I do not see how this is to be got out of the Chinese. Chang Yü gives a clear paraphrase: 士卒死戰安不得志, and quotes his favourite Wei Liao Tzŭ (ch. 3): 一夫仗劍擊於市萬人無不避之者臣謂非一人之獨勇萬人皆不肖也何則必死與必生固不侔也 "If one man were to run amok with a sword in the market-place, and everybody else tried to get out of his way, I should not allow that this man alone had courage and that all the rest were contemptible cowards. The truth is, that a desperado and a man who sets some value on his life do not meet on even terms."

Officers and men alike will put forth their uttermost strength.

士人 appears to stand for the more usual 士卒. Chang Yü says: 同在難地安得不共竭其力 "If they are in an awkward place together, they will surely exert their united strength to get out of it."

24. Soldiers when in desperate straits lose the sense of fear. If there is no place of refuge, they will stand firm. If they are in the heart of a hostile country, they will show a stubborn front.

Capt. Calthrop weakly says: "there is unity," as though the text were 則專, as in § 20. But 拘 introduces quite a new idea — that of *tenacity* — which Ts'ao Kung tries to explain by the word 縛 "to bind fast."

If there is no help for it, they will fight hard.

25. 是故其兵不修而戒不求而得不約而親不
令而信

26. 禁祥去疑至死無所災

25. Thus, without waiting to be marshalled, the soldiers
will be constantly on the *qui vive*;

Tu Mu says: 不待修整而自戒懼. Capt. Calthrop wrongly
translates 不修 "without warnings."

without waiting to be asked, they will do your will;

Literally, "without asking, you will get." Chang Yü's paraphrase is:
不求索而得情意.

without restrictions, they will be faithful;

Chang Yü says: 不約束而親上.

without giving orders, they can be trusted.

This last clause is very similar in sense to the one preceding, except
that 親 indicates the soldiers' attachment to their leader, and 信 the
leader's attitude towards them. I rather doubt if 信 can mean "they
will have confidence in their leader," as the commentary seems to indicate. That way, the sense is not nearly so good. On the other hand, it
is just possible that here, as in VIII. § 8 and *infra*, § 55, 信 may = 申:
"without orders, they will carry out [their leader's plans]." The whole
of this paragraph, of course, has reference to "desperate ground."

26. Prohibit the taking of omens, and do away with
superstitious doubts.

祥 is amplified by Ts'ao Kung into 妖祥之言, and 疑 into
疑惑之計. Cf. the *Ssŭ-ma Fa*, ch. 3: 滅厲祥.

Then, until death itself comes, no calamity need be feared.

The superstitious, "bound in to saucy doubts and fears," degenerate
into cowards and "die many times before their deaths." Tu Mu quotes
Huang Shih-kung: 禁巫祝不得為吏士卜問軍之吉
凶恐亂軍士之心 "'Spells and incantations should be strictly
forbidden, and no officer allowed to inquire by divination into the fortunes
of an army, for fear the soldier's minds should be seriously perturbed.'
The meaning is," he continues, "that if all doubts and scruples are dis-

27. 吾士無餘財非惡貨也無餘命非惡壽也

28. 令發之日士卒坐者涕霑襟偃臥者涕交頤
投之無所往者諸劌之勇也

carded, your men will never falter in their resolution until they die."
The reading of the standard text is 無所之 "there will be no refuge,"
which does not fit in well here. I therefore prefer to adopt the variant
災, which evidently stood in Li Ch'üan's text.

27. If our soldiers are not overburdened with money,
it is not because they have a distaste for riches; if their
lives are not unduly long, it is not because they are dis-
inclined to longevity.

Chang Yü has the best note on this passage: 貨與壽人之所
愛也所以燒擲財寶割棄性命者非憎惡之也
不得已也 "Wealth and long life are things for which all men have
a natural inclination. Hence, if they burn or fling away valuables, and
sacrifice their own lives, it is not that they dislike them, but simply that
they have no choice." Sun Tzŭ is slyly insinuating that, as soldiers are
but human, it is for the general to see that temptations to shirk fighting
and grow rich are not thrown in their way. Capt. Calthrop, mistaking
惡 for the adjective, has: "not because money is a bad thing ... not
because long life is evil."

28. On the day they are ordered out to battle, your
soldiers may weep,

The word in the Chinese is 涕 "snivel." This is taken to indicate
more genuine grief than tears alone.

those sitting up bedewing their garments, and those lying
down letting the tears run down their cheeks.

Not because they are afraid, but because, as Ts'ao Kung says, 皆持必
死之計 "all have embraced the firm resolution to do or die." We
may remember that the heroes of the Iliad were equally childlike in
showing their emotion. Chang Yü alludes to the mournful parting at the
易 I River between 荊軻 Ching K'o and his friends, when the former
was sent to attempt the life of the King of Ch'in (afterwards First Emperor)
in 227 B.C. The tears of all flowed down like rain as he bade them
farewell and uttered the following lines: 風蕭蕭兮、易水寒、

29. 故善用兵譬如率然率然者常山之虵也擊
其首則尾至擊其尾則首至擊其中則首尾
俱至

壯士一去兮、不復還 "The shrill blast is blowing, Chilly the burn; Your champion is going — Not to return."*
But let them once be brought to bay, and they will display the courage of a Chu or a Kuei.

諸 was the personal name of 專諸 Chuan Chu, a native of the Wu State and contemporary with Sun Tzŭ himself, who was employed by 公子光 Kung-tzŭ Kuang, better known as Ho Lü Wang, to assassinate his sovereign 王僚 Wang Liao with a dagger which he secreted in the belly of a fish served up at a banquet. He succeeded in his attempt, but was immediately hacked to pieces by the king's bodyguard. This was in 515 B.C. The other hero referred to, 曹歲 Ts'ao Kuei (or Ts'ao 沫 Mo), performed the exploit which has made his name famous 166 years earlier, in 681 B.C. Lu had been thrice defeated by Ch'i, and was just about to conclude a treaty surrendering a large slice of territory, when Ts'ao Kuei suddenly seized 桓公 Huan Kung, the Duke of Ch'i, as he stood on the altar steps and held a dagger against his chest. None of the Duke's retainers dared to move a muscle, and Ts'ao Kuei proceeded to demand full restitution, declaring that Lu was being unjustly treated because she was a smaller and weaker state. Huan Kung, in peril of his life, was obliged to consent, whereupon Ts'ao Kuei flung away his dagger and quietly resumed his place amid the terrified assemblage without having so much as changed colour. As was to be expected, the Duke wanted afterwards to repudiate the bargain, but his wise old counsellor 管仲 Kuan Chung pointed out to him the impolicy of breaking his word, and the upshot was that this bold stroke regained for Lu the whole of what she had lost in three pitched battles. [For another anecdote of Ts'ao Kuei see VII. § 27, note; and for the biographies of these three bravos, Ts'ao, Chuan and Ching, see *Shih Chi*, ch. 86.]

29. The skilful tactician may be likened to the *shuai-jan*. Now the *shuai-jan* is a snake that is found in the Ch'ang mountains.

率然 means "suddenly" or "rapidly," and the snake in question was doubtless so called owing to the rapidity of its movements. Through this passage, the term has now come to be used in the sense of "military manœuvres." The 常山 have apparently not been identified.

---

* Giles' Biographical Dictionary, no. 399.

30. 敢問兵可使如率然乎曰可夫吳人與越人
    相惡也當其同舟而濟遇風其相救也如左
    右手

31. 是故方馬埋輪未足恃也

Strike at its head, and you will be attacked by its tail;
strike at its tail, and you will be attacked by its head;
strike at its middle,

Another reading in the *Yü Lan* for 中 is 腹 "belly."

and you will be attacked by head and tail both.

30. Asked if an army can be made to imitate the
shuai-jan,

That is, as Mei-Yao-ch'ên says, 可使兵首尾率然相應
如一體乎 "Is it possible to make the front and rear of an army
each swiftly responsive to attack on the other, just as though they were
parts of a single living body?"

I should answer, Yes. For the men of Wu and the men
of Yüeh are enemies;

Cf. VI. § 21.

yet if they are crossing a river in the same boat and
are caught by a storm, they will come to each other's
assistance just as the left hand helps the right.

The meaning is: If two enemies will help each other in a time of common peril, how much more should two parts of the same army, bound
together as they are by every tie of interest and fellow-feeling. Yet it is
notorious that many a campaign has been ruined through lack of co-
operation, especially in the case of allied armies.

31. Hence it is not enough to put one's trust in the
tethering of horses,

方 is said here to be equivalent to 縛.

and the burying of chariot wheels in the ground.

These quaint devices to prevent one's army from running away recall
the Athenian hero Sôphanes, who carried an anchor with him at the
battle of Plataea, by means of which he fastened himself firmly to one
spot. [See Herodotus, IX. 74.] It is not enough, says Sun Tzŭ, to render
flight impossible by such mechanical means. You will not succeed unless

9

32. 齊勇若一政之道也
33. 剛柔皆得地之理也
34. 故善用兵者攜手若使一人不得已也

your men have tenacity and unity of purpose, and, above all, a spirit of sympathetic co-operation. This is the lesson which can be learned from the *shuai-jan*.

32. The principle on which to manage an army is to set up one standard of courage which all must reach.

Literally, "level the courage [of all] as though [it were that of] one." If the ideal army is to form a single organic whole, then it follows that the resolution and spirit of its component parts must be of the same quality, or at any rate must not fall below a certain standard. Wellington's seemingly ungrateful description of his army at Waterloo as "the worst he had ever commanded" meant no more than that it was deficient in this important particular — unity of spirit and courage. Had he not foreseen the Belgian defections and carefully kept those troops in the background, he would almost certainly have lost the day.

33. How to make the best of both strong and weak — that is a question involving the proper use of ground.

This is rather a hard sentence on the first reading, but the key to it will be found, firstly, in the pause after 得, and next, in the meaning of 得 itself. The best equivalent for this that I can think of is the German "zur Geltung kommen." Mei Yao-ch'ên's paraphrase is: 兵無強弱皆得用者是因地之勢也 "The way to eliminate the differences of strong and weak and to make both serviceable is to utilise accidental features of the ground." Less reliable troops, if posted in strong positions, will hold out as long as better troops on more exposed terrain. The advantage of position neutralises the inferiority in stamina and courage. Col. Henderson says: "With all respect to the text books, and to ordinary tactical teaching, I am inclined to think that the study of ground is often overlooked, and that by no means sufficient importance is attached to the selection of positions ... and to the immense advantages that are to be derived, whether you are defending or attacking, from the proper utilisation of natural features." *

34. Thus the skilful general conducts his army just as though he were leading a single man, willy-nilly, by the hand.

---

* "The Science of War," p. 333.

35. 將軍之事靜以幽正以治
36. 能愚士卒之耳目 使之無知

Tu Mu says: 喻易也 "The simile has reference to the ease with which he does it." 不得已 means that he makes it impossible for his troops to do otherwise than obey. Chang Yü quotes a jingle, to be found in Wu Tzǔ, ch. 4: 將之所揮、莫不從移、將之所指、莫不前死.

35. It is the business of a general to be quiet and thus ensure secrecy; upright and just, and thus maintain order.

靜 seems to combine the meanings "noiseless" and "imperturbable," both of which attributes would of course conduce to secrecy. Tu Mu explains 幽 as 幽深難測 "deep and inscrutable," and 正 as 平正無偏 "fair and unbiassed." Mei Yao-ch'ên alone among the commentators takes 治 in the sense of 自治 "self-controlled." 幽 and 治 are causally connected with 靜 and 正 respectively. This is not brought out at all in Capt. Calthrop's rendering: "The general should be calm, inscrutable, just and prudent." The last adjective, moreover, can in no sense be said to represent 治.

36. He must be able to mystify his officers and men by false reports and appearances,

Literally, "to deceive their eyes and ears" — 愚 being here used as a verb in the sense of 誤.

and thus keep them in total ignorance.

Ts'ao Kung gives us one of his excellent apophthegms: 民可與樂成不可與慮始 "The troops must not be allowed to share your schemes in the beginning; they may only rejoice with you over their happy outcome." "To mystify, mislead, and surprise the enemy," is one of the first principles in war, as has been frequently pointed out. But how about the other process — the mystification of one's own men? Those who may think that Sun Tzǔ is over-emphatic on this point would do well to read Col. Henderson's remarks on Stonewall Jackson's Valley campaign: "The infinite pains," he says, "with which Jackson sought to conceal, even from his most trusted staff officers, his movements, his intentions, and his thoughts, a commander less thorough would have pronounced useless" — etc. etc.* In the year 88 A.D., as we read in ch. 47

---

* "Stonewall Jackson," vol. I, p. 421.

37. 易其事革其謀使人無識易其居迂其途使
人不得慮

of the *Hou Han Shu*, "Pan Ch'ao took the field with 25,000 men from Khotan and other Central Asian states with the object of crushing Yarkand. The King of Kutcha replied by dispatching his chief commander to succour the place with an army drawn from the kingdoms of Wên-su, Kumo and Wei-t'ou, totalling 50,000 men. Pan Ch'ao summoned his officers and also the King of Khotan to a council of war, and said: 'Our forces are now outnumbered and unable to make head against the enemy. The best plan, then, is for us to separate and disperse, each in a different direction. The King of Khotan will march away by the easterly route, and I will then return myself towards the west. Let us wait until the evening drum has sounded and then start.' Pan Ch'ao now secretly released the prisoners whom he had taken alive, and the King of Kutcha was thus informed of his plans. Much elated by the news, the latter set off at once at the head of 10,000 horsemen to bar Pan Ch'ao's retreat in the west, while the King of Wên-su rode eastwards with 8000 horse in order to intercept the King of Khotan. As soon as Pan Ch'ao knew that the two chieftains had gone, he called his divisions together, got them well in hand, and at cock-crow hurled them against the army of Yarkand, as it lay encamped. The barbarians, panic-stricken, fled in confusion, and were closely pursued by Pan Ch'ao. Over 5000 heads were brought back as trophies, besides immense spoils in the shape of horses and cattle and valuables of every description. Yarkand then capitulating, Kutcha and the other kingdoms drew off their respective forces. From that time forward, Pan Ch'ao's prestige completely overawed the countries of the west." In this case, we see that the Chinese general not only kept his own officers in ignorance of his real plans, but actually took the bold step of dividing his army in order to deceive the enemy.

37. By altering his arrangements and changing his plans,

Wang Hsi thinks that this means, not using the same stratagem twice. He says: 已行之事已施之謀當革易之不可再之.

he keeps the enemy without definite knowledge.

Note that 人 denotes the *enemy*, as opposed to the 士卒 of § 36. Capt. Calthrop, not perceiving this, joins the two paragraphs into one. Chang Yü quotes 太白山人 as saying: 兵貴詭道者非止詭敵也抑詭我士卒使由而不使知之也 "The axiom, that war is based on deception, does not apply only to deception of the enemy. You must deceive even your own soldiers. Make them follow you, but without letting them know why."

38. 帥 與 之 期 如 登 高 而 去 其 梯 帥 與 之 深 入 諸
侯 之 地 而 發 其 機

39. 焚 舟 破 釜 若 驅 羣 羊 而 往 驅 而 來 莫 知 所 之

By shifting his camp and taking circuitous routes, he prevents the enemy from anticipating his purpose.

Wang Hsi paraphrases 易 其 居 as 處 易 者 "camp on easy ground," and Chang Yü follows him, saying: 其 居 則 去 險 而 就 易. But this is an utterly untenable view. For 迂 其 途, cf. VII. 4. Chia Lin, retaining his old interpretation of those words, is now obliged to explain 易 其 居 as "cause the enemy to shift his camp," which is awkward in the extreme.

38. At the critical moment, the leader of an army acts like one who has climbed up a height and then kicks away the ladder behind him.

I must candidly confess that I do not understand the syntax of 帥 與 之 期, though the meaning is fairly plain. The difficulty has evidently been felt, for Tu Mu tells us that one text omits 期 如. It is more likely, however, that a couple of characters have dropped out.

He carries his men deep into hostile territory before he shows his hand.

發 其 機, literally, "releases the spring" (see V. § 15), that is, takes some decisive step which makes it impossible for the army to return — like 項 羽 Hsiang Yü, who sunk his ships after crossing a river. Ch'ên Hao, followed by Chia Lin, understands the words less well as 發 其 心 機 "puts forth every artifice at his command." But 機 in this derived sense occurs nowhere else in Sun Tzŭ.

39. He burns his boats and breaks his cooking-pots;

Omitted in the *T'u Shu*.

like a shepherd driving a flock of sheep, he drives his men this way and that, and none knows whither he is going.

The *T'u Shu* inserts another 驅 after 羊. Tu Mu says: 三 軍 但 知 進 退 之 命 不 知 攻 取 之 端 也 "The army is only cognisant of orders to advance or retreat; it is ignorant of the ulterior ends of attacking and conquering."

40. 聚三軍之眾投之於險此謂將軍之事也
41. 九地之變屈伸之利人情之理不可不察也
42. 凡爲客之道深則專淺則散
43. 去國越境而師者絶地也四達者衢地也

40. To muster his host and bring it into danger: —
this may be termed the business of the general.

Sun Tzŭ means that after mobilisation there should be no delay in
aiming a blow at the enemy's heart. With 投之於險 cf. *supra*,
§ 23: 投之無所往. Note how he returns again and again to this
point. Among the warring states of ancient China, desertion was no
doubt a much more present fear and serious evil than it is in the armies
of to-day.

41. The different measures suited to the nine varieties
of ground;

Chang Yü says: 九地之法不可拘泥 "One must not be
hide-bound in interpreting the rules for the nine varieties of ground."

the expediency of aggressive or defensive tactics;

The use of 屈伸 "contraction and expansion" may be illustrated by
the saying 屈以求伸, which almost exactly corresponds to the French
"il faut reculer pour mieux sauter." * Capt. Calthrop, *more suo*, avoids
a real translation and has: "the suiting of the means to the occasion."

and the fundamental laws of human nature: these are
things that must most certainly be studied.

42. When invading hostile territory, the general principle is, that penetrating deeply brings cohesion; penetrating
but a short way means dispersion.

Cf. *supra*, § 20.

43. When you leave your own country behind, and
take your army across neighbouring territory,

Chang Yü's paraphrase is 而用師者.

you find yourself on critical ground.

---

* See Giles' Dictionary, no. 9817.

44. 入深者重地也入淺者輕地也

45. 背固前隘者圍地也無所往者死地也

46. 是故散地吾將一其志輕地吾將使之屬

This "ground" is cursorily mentioned in VIII. § 2, but it does not figure among the Nine 地 of this chapter or the Six 地形 in chap. X. One's first impulse would be to translate it "distant ground" (絕域 is commonly used in the sense of "distant lands"), but this, if we can trust the commentators, is precisely what is not meant here. Mei Yao-ch'ên says it is 進不及輕退不及散在二地之間也 "a position not far enough advanced to be called 'facile,' and not near enough to home to be called 'dispersive,' but something between the two." That, of course, does not explain the name 絕, which seems to imply that the general has severed his communications and temporarily cut himself off from his base. Thus, Wang Hsi says: "It is ground separated from home by an interjacent state, whose territory we have had to cross in order to reach it. Hence it is incumbent on us to settle our business there quickly." He adds that this position is of rare occurrence, which is the reason why it is not included among the 九地. Capt. Calthrop gives but a poor rendering of this sentence: "To leave home and cross the borders is to be free from interference."

When there are means of communication

The *T'u Shu* reads 通 for 達.

on all four sides, the ground is one of intersecting highways.

From 四達 down to the end of § 45, we have some of the definitions of the early part of the chapter repeated in slightly different language. Capt. Calthrop omits these altogether.

44. When you penetrate deeply into a country, it is serious ground. When you penetrate but a little way, it is facile ground.

45. When you have the enemy's strongholds on your rear,

固 = 險固.

and narrow passes in front, it is hemmed-in ground. When there is no place of refuge at all, it is desperate ground.

46. Therefore, on dispersive ground, I would inspire my men with unity of purpose.

47. 爭地吾將趨其後

This end, according to Tu Mu, is best attained by remaining on the defensive, and avoiding battle. Cf. *supra*, § 11.

On facile ground, I would see that there is close connection between all parts of my army.

The *T'ung Tien* has 其 instead of 之. The present reading is supported by the 遺說 of Chêng Yu-hsien. As Tu Mu says, the object is to guard against two possible contingencies: 一者備其逃逸 二者恐其敵至 "(1) the desertion of our own troops; (2) a sudden attack on the part of the enemy." Cf. VII. § 17: 其徐如林. Mei Yao-ch'ên says: 行則隊校相繼止則營壘聯屬 "On the march, the regiments should be in close touch; in an encampment, there should be continuity between the fortifications." He seems to have forgotten, by the way, what Sun Tzŭ says above: 輕地則無止.

47. On contentious ground, I would hurry up my rear.

This is Ts'ao Kung's interpretation. Chang Yü adopts its, saying: 當疾進其後使首尾俱至 "We must quickly bring up our rear, so that head and tail may both reach the goal." That is, they must not be allowed to straggle up a long way apart. Mei Yao-ch'ên offers another equally plausible explanation: 敵未至其地我若在後則當疾趨以爭之 "Supposing the enemy has not yet reached the coveted position, and we are behind him, we should advance with all speed in order to dispute its possession." 其 would thus denote the enemy, 後 being the preposition, and 趨 would retain its usual intransitive sense. Cf. VII. § 4: 後人發先人至. Ch'ên Hao, on the other hand, assuming that the enemy has had time select his own ground, quotes VI. § 1, where Sun Tzŭ warns us against coming exhausted to the attack. His own idea of the situation is rather vaguely expressed: 若地利在前先分精銳以據之彼若恃眾來爭我以大眾趨其後無不尅者 "If there is a favourable position lying in front of you, detach a picked body of troops to occupy it; then if the enemy, relying on their numbers, come up to make a fight for it, *you may fall quickly on their rear* with your main body, and victory will be assured." It was thus, he adds, that Chao Shê beat the army of Ch'in. [See p. 57.] Li Ch'üan would read 多 for 趨, it is not easy to see why.

48. 交地吾將謹其守衢地吾將固其結
49. 重地吾將繼其食圮地吾將進其塗
50. 圍地吾將塞其闕死地吾將示之以不活

48. On open ground, I would keep a vigilant eye on my defences.

As Wang Hsi says, 懼襲我也 "fearing a surprise attack." The *T'ung Tien* reads here 固其結 (see next sentence).

On ground of intersecting highways, I would consolidate my alliances.

The *T'ung Tien* reads 謹其市, which Tu Yu explains as "watching the market towns," 變事之端 "the hotbeds of revolution." Capt. Calthrop translates 固其結 by the same words as 合交 in § 12: "cultivate intercourse."

49. On serious ground, I would try to ensure a continuous stream of supplies.

The commentators take this as referring to forage and plunder, not, as one might expect, to an unbroken communication with a home base. One text, indeed, gives the reading 掠其食. Cf. § 13. Capt. Calthrop's "be careful of supplies" fails to render the force of 繼.

On difficult ground, I would keep pushing on along the road.

Capt. Calthrop's "do not linger" cannot be called a translation, but only a paraphrase of the paraphrase offered by Ts'ao Kung: 疾過去也 "Pass away from it in all haste."

50. On hemmed-in ground, I would block any way of retreat.

意欲突圍示以守固 "To make it seem that I mean to defend the position, whereas my real intention is to burst suddenly through the enemy's lines" [Mêng Shih]; 使士卒必死戰也 "in order to make my soldiers fight with desperation" [Mei Yao-ch'ên]; 懼人有走心 "fearing lest my men be tempted to run away" [Wang Hsi]. Tu Mu points out that this is the converse of VII. § 36, where it is the enemy who is surrounded. In 532 A.D., 高歡 Kao Huan, afterwards Emperor and canonised as 神武 Shên-wu, was surrounded by a great army under 爾朱兆 Êrh-chu Chao and others. His own force was comparatively small, consisting only of 2000 horse and something under 30,000 foot. The lines of investment had not been drawn very closely together, gaps being left at certain points. But Kao Huan, instead of trying to escape, actually made a shift to block all the remaining outlets himself by driving into them a number of oxen and donkeys roped together. As soon as his officers and men saw that there was nothing for it but to conquer or die, their spirits rose to an extraordinary pitch of exaltation, and they charged with such desperate ferocity that the opposing ranks broke and crumbled under their onslaught. [See Tu Mu's commentary, and 北齊書 ch. 1, fol. 6.]

On desperate ground, I would proclaim to my soldiers the hopelessness of saving their lives.

Tu Yu says: 焚輜重棄糧食塞井夷竈示之無活 必殊死戰也 "Burn your baggage and impedimenta, throw away your stores and provisions, choke up the wells, destroy your cooking-stoves, and make it plain to your men that they cannot survive, but must fight to the death." Mei Yao-ch'ên says epigrammatically: 必死 可生 "The only chance of life lies in giving up all hope of it." This concludes what Sun Tzŭ has to say about "grounds" and the "variations" corresponding to them. Reviewing the passages which bear on this important subject, we cannot fail to be struck by the desultory and unmethodical fashion in which it is treated. Sun Tzŭ begins abruptly in VIII. § 2 to enumerate "variations" before touching on "grounds" at all, but only mentions five, namely nos. 7, 5, 8 and 9 of the subsequent list, and one that is not included in it. A few varieties of ground are dealt with in the earlier portion of chap. IX, and then chap. X sets forth six new grounds, with six variations of plan to match. None of these is mentioned again, though the first is hardly to be distinguished from ground no. 4 in the next chapter. At last, in chap. XI, we come to the Nine Grounds *par excellence*, immediately followed by the variations. This takes us down to § 14. In §§ 43—45, fresh definitions are provided for nos. 5, 6, 2, 8 and 9 (in the order given), as well as for the tenth ground noticed in chap VIII; and finally, the nine variations are enumerated once more from beginning to end, all, with the exception of 5, 6 and 7, being different from those previously given. Though it is impossible to account for the present state of Sun Tzŭ's text, a few suggestive facts may be brought into prominence: (1) Chap. VIII, according to the title, should deal with nine variations, whereas only five appear. (2) It is an abnormally short chapter. (3) Chap. XI is entitled The Nine Grounds. Several of these are defined twice over, besides which there are two distinct lists of the corresponding variations. (4) The length of the chapter is disproportionate, being double that of any other except IX. I do not propose

51. 故兵之情圍則禦不得已則鬥過則從

to draw any inferences from these facts, beyond the general conclusion that Sun Tzŭ's work cannot have come down to us in the shape in which it left his hands: chap. VIII is obviously defective and probably out of place, while XI seems to contain matter that has either been added by a later hand or ought to appear elsewhere.

51. For it is the soldier's disposition to offer an obstinate resistance when surrounded, to fight hard when he cannot help himself, and to obey promptly when he has fallen into danger.

過則從 is rendered by Capt. Calthrop: "to pursue the enemy if he retreat." But 過 cannot mean "to retreat." Its primary sense is to pass over, hence to go too far, to exceed or to err. Here, however, the word has lost all implication of censure, and appears to mean "to pass the boundary line dividing safety from danger," or, as Chang Yü puts it, 深陷于危難之地 "to be deeply involved in a perilous position." The latter commentator alludes to the conduct of Pan Ch'ao's devoted followers in 73 A.D. The story runs thus in the *Hou Han Shu*, ch. 47, fol. 1 v°: "When Pan Ch'ao arrived at 鄯善 Shan-shan, 廣 Kuang, the King of the country, received him at first with great politeness and respect; but shortly afterwards his behaviour underwent a sudden change, and he became remiss and negligent. Pan Ch'ao spoke about this to the officers of his suite: 'Have you not noticed,' he said, 'that Kuang's polite intentions are on the wane? This must signify that envoys have come from the Northern barbarians, and that consequently he is in a state of indecision, not knowing with which side to throw in his lot. That surely is the reason. The truly wise man, we are told, can perceive things before they have come to pass; how much more, then, those that are already manifest!' Thereupon he called one of the natives who had been assigned to his service, and set a trap for him, saying: 'Where are those envoys from the Hsiung-nu who arrived some days ago?' The man was so taken aback that between surprise and fear he presently blurted out the whole truth. Pan Ch'ao, keeping his informant carefully under lock and key, then summoned a general gathering of his officers, thirty-six in all, and began drinking with them. When the wine had mounted into their heads a little, he tried to rouse their spirit still further by addressing them thus: 'Gentlemen, here we are in the heart of an isolated region, anxious to achieve riches and honour by some great exploit. Now it happens that an ambassador from the Hsiung-nu arrived in this kingdom only a few days ago, and the result is that the respectful courtesy extended towards us by our royal host has disappeared. Should this envoy prevail upon him to seize our party and hand us over to the Hsiung-nu,

52. 是故不知諸侯之謀者不能預交不知山林
險阻沮澤之形者不能行軍不用鄉導者不
能得地利

53. 四五者不知一非霸王之兵也

our bones will become food for the wolves of the desert. What are we
to do?' With one accord, the officers replied: '*Standing as we do in peril
of our lives, we will follow our commander through life and death*'
（今在危亡之地死生從司馬）." For the sequel of this
adventure, see chap. XII. § 1, note.

52. We cannot enter into alliance with neighbouring
princes until we are acquainted with their designs. We
are not fit to lead an army on the march unless we are
familiar with the face of the country — its mountains and
forests, its pitfalls and precipices, its marshes and swamps.
We shall be unable to turn natural advantages to account
unless we make use of local guides.

These three sentences are repeated from VII. §§ 12—14 — in order to
emphasise their importance, the commentators seem to think. I prefer
to regard them as interpolated here in order to form an antecedent to
the following words. With regard to local guides, Sun Tzŭ might have
added that there is always the risk of going wrong, either through their
treachery or some misunderstanding such as Livy records (XXII. 13):
Hannibal, we are told, ordered a guide to lead him into the neighbourhood
of Casinum, where there was an important pass to be occupied; but his
Carthaginian accent, unsuited to the pronunciation of Latin names, caused
the guide to understand Casilinum instead of Casinum, and turning from
his proper route, he took the army in that direction, the mistake not
being discovered until they had almost arrived.

53. To be ignorant of any one of the following four
or five principles

Referring, I think, to what is contained in §§ 54, 55. Ts'ao Kung,
thinking perhaps of the 五利 in VIII. § 6, takes them to be 九地
之利害 "the advantages and disadvantages attendant on the nine
varieties of ground." The *T'u Shu* reads 此五者.

does not befit a warlike prince.

霸王, "one who rules by force," was a term specially used for those
princes who established their hegemony over other feudal states. The

54. 夫霸王之兵伐大國則其眾不得聚威加於
敵則其交不得合

55. 是故不爭天下之交不養天下之權信己之
私威加於敵故其城可拔其國可隳

famous 五霸 of the 7th century B.C. were (1) 齊桓公 Duke Huan
of Ch'i (2) 晉文公 Duke Wên of Chin, (3) 宋襄公 Duke Hsiang
of Sung, (4) 楚莊王 Prince Chuang of Ch'u, (5) 秦穆公 Duke
Mu of Ch'in. Their reigns covered the period 685—591 B.C.

54. When a warlike prince attacks a powerful state,
his generalship shows itself in preventing the concentration
of the enemy's forces. He overawes his opponents,

Here and in the next sentence, the *Yü Lan* inserts 家 after 敵.

and their allies are prevented from joining against him.

Mei Yao-ch'ên constructs one of the chains of reasoning that are so
much affected by the Chinese: "In attacking a powerful state, if you can
divide her forces, you will have a superiority in strength; if you have a
superiority in strength, you will overawe the enemy; if you overawe the
enemy, the neighbouring states will be frightened; and if the neighbouring
states are frightened, the enemy's allies will be prevented from joining
her." The following gives a stronger meaning to 威加: 若大國
一敗則小國離而不聚矣 "If the great state has once been
defeated (before she has had time to summon her allies), then the lesser
states will hold aloof and refrain from massing their forces." Ch'ên Hao
and Chang Yü take the sentence in quite another way. The former
says: "Powerful though a prince may be, if he attacks a large state, he
will be unable to raise enough troops, and must rely to some extent on
external aid; if he dispenses with this, and with overweening confidence
in his own strength, simply tries to intimidate the enemy, he will surely
be defeated." Chang Yü puts his view thus: "If we recklessly attack a
large state, our own people will be discontented and hang back. But if
(as will then be the case) our display of military force is inferior by half
to that of the enemy, the other chieftains will take fright and refuse to
join us." According to this interpretation, 其 would refer, not to the
大國, but to the 霸王 himself.

55. Hence he does not strive

For 爭 the *Yü Lan* reads 事.

56. 施無法之賞懸無政之令犯三軍之眾若使一人

to ally himself with all and sundry,

天下, as in § 6, stands for 諸侯 "the feudal princes," or the states ruled by them.

nor does he foster the power of other states. He carries out his own secret designs,

For 信 (read *shên*[1]) in the meaning of 伸, cf. VIII. § 8. The commentators are unanimous on this point, and we must therefore beware of translating 信己之私 by "secretly self-confident" or the like. Capt. Calthrop (omitting 之私) has: "he has confidence in himself."

keeping his antagonists in awe.

The train of thought appears to be this: Secure against a combination of his enemies, 能絕天下之交惟得伸己之私志威而無外交者 "he can afford to reject entangling alliances and simply pursue his own secret designs, his prestige enabling him to dispense with external friendships." (Li Ch'üan.)

Thus he is able to capture their cities and overthrow their kingdoms.

This paragraph, though written many years before the Ch'in State became a serious menace, is not a bad summary of the policy by which the famous Six Chancellors gradually paved the way for her final triumph under Shih Huang Ti. Chang Yü, following up his previous note, thinks that Sun Tzŭ is condemning this attitude of cold-blooded selfishness and haughty isolation. He again refers 其 to the warlike prince, thus making it appear that in the end he is bound to succumb.

56. Bestow rewards without regard to rule,

Wu Tzŭ (ch. 3) less wisely says: 進有重賞退有重刑 "Let advance be richly rewarded and retreat be heavily punished."

issue orders

懸, literally, "hang" or "post up."

without regard to previous arrangements;

杜姦婾 "In order to prevent treachery," says Wang Hsi. The general meaning is made clear by Ts'ao Kung's quotation from the

57. 犯之以事勿告以言犯之以利勿告以害
58. 投之亡地然後存陷之死地然後生

Ssŭ-ma Fa: 見敵作誓瞻功作賞 "Give instructions only on sighting the enemy; give rewards only when you see deserving deeds." 無政, however, presents some difficulty. Ts'ao Kung's paraphrase, 軍法令不應預施懸也, I take to mean: "The final instructions you give to your army should not correspond with those that have been previously posted up." Chang Yü simplifies this into 政不預告 "your arrangements should not be divulged beforehand." And Chia Lin says: 不守常法常政 "there should be no fixity in your rules and arrangements." Not only is there danger in letting your plans be known, but war often necessitates the entire reversal of them at the last moment.

and you will be able to handle a whole army

犯, according to Ts'ao Kung, is here equal to 用. The exact meaning is brought out more clearly in the next paragraph.

as though you had to do with but a single man.

Cf. supra, § 34.

57. Confront your soldiers with the deed itself; never let them know your design.

Literally, "do not tell them words;" *i.e.* do not give your reasons for any order. Lord Mansfield once told a junior colleague to "give no reasons" for his decisions, and the maxim is even more applicable to a general than to a judge. Capt. Calthrop translates this sentence with beautiful simplicity: "Orders should direct the soldiers." That is all.

When the outlook is bright, bring it before their eyes; but tell them nothing when the situation is gloomy.

58. Place your army in deadly peril, and it will survive; plunge it into desperate straits, and it will come off in safety.

Compare the paradoxical saying 亡者存之基死者生之本. These words of Sun Tzŭ were once quoted by Han Hsin in explanation of the tactics he employed in one of his most brilliant battles, already alluded to on p. 28. In 204 B.C., he was sent against the army of Chao, and halted ten miles from the mouth of the 井陘 Chinghsing pass, where the enemy had mustered in full force. Here, at midnight, he detached a body of 2000 light cavalry, every man of which was furnished with a red flag. Their instructions were to make their way through narrow defiles and keep a secret watch on the enemy. "When the men of Chao see me in full flight," Han Hsin said, "they will abandon their fortifications and give chase. This must be the sign for you to rush in, pluck down the Chao standards and set up the red banners of 漢 Han in their stead." Turning then to his other officers, he remarked: "Our adversary holds a strong position, and is not likely to come out and attack us until he sees the standard and drums of the commander-in-chief, for fear I should turn back and escape through the mountains." So saying, he first of all sent out a division consisting of 10,000 men, and ordered them to form in line of battle with their backs to the River 泜 Ti. Seeing this manœuvre, the whole army of Chao broke into loud laughter. By this time it was broad daylight, and Han Hsin, displaying the generalissimo's flag, marched out of the pass with drums beating, and was immediately engaged by the enemy. A great battle followed, lasting for some time; until at length Han Hsin and his colleague 張耳 Chang Ni, leaving drums and banner on the field, fled to the division on the river bank, where another fierce battle was raging. The enemy rushed out to pursue them and to secure the trophies, thus denuding their ramparts of men; but the two generals succeeded in joining the other army, which was fighting with the utmost desperation. The time had now come for the 2000 horsemen to play their part. As soon as they saw the men of Chao following up their advantage, they galloped behind the deserted walls, tore up the enemy's flags and replaced them by those of Han. When the Chao army turned back from the pursuit, the sight of these red flags struck them with terror. Convinced that the Hans had got in and overpowered their king, they broke up in wild disorder, every effort of their leader to stay the panic being in vain. Then the Han army fell on them from both sides and completed the rout, killing a great number and capturing the rest, amongst whom was King 歇 Ya himself . . . . After the battle, some of Han Hsin's officers came to him and said: "In the *Art of War* we are told to have a hill or tumulus on the right rear, and a river or marsh on the left front. [This appears to be a blend of Sun Tzŭ and T'ai Kung. See IX. § 9, and note.] You, on the contrary, ordered us to draw up our troops with the river at our back. Under these conditions, how did you manage to gain the victory?" The general replied: "I fear you gentlemen have not studied the Art of War with sufficient care. Is it not written there: '*Plunge your army into desperate straits and it will come off in safety; place it in deadly peril and it will survive*'? Had I taken the usual course, I should never have been able to bring my colleagues round. What says the Military Classic (經)? — 'Swoop down on the market-place and drive the men off to fight' (毆市人而戰之). [This passage does not occur in the present text of Sun Tzŭ.] If I had not placed my troops in a position where

59. 夫眾陷於害然後能爲勝敗
60. 故爲兵之事在於順詳敵之意
61. 并敵一向千里殺將
62. 此謂巧能成事者也

they were obliged to fight for their lives, but had allowed each man to follow his own discretion, there would have been a general *débandade*, and it would have been impossible to do anything with them." The officers admitted the force of his argument, and said: "These are higher tactics than we should have been capable of." [See *Ch'ien Han Shu*, ch. 34, ff. 4, 5.]

59. For it is precisely when a force has fallen into harm's way that it is capable of striking a blow for victory.

Danger has a bracing effect.

60. Success in warfare is gained by carefully accommodating ourselves to the enemy's purpose.

Ts'ao Kung says: 佯愚也 "Feign stupidity" — by an appearance of yielding and falling in with the enemy's wishes. Chang Yü's note makes the meaning clear: "If the enemy shows an inclination to advance, lure him on to do so; if he is anxious to retreat, delay on purpose that he may carry out his intention." The object is to make him remiss and contemptuous before we deliver our attack.

61. By persistently hanging on the enemy's flank,

I understand the first four words to mean "accompanying the enemy in one direction." Ts'ao Kung says: 并兵向敵 "unite the soldiers and make for the enemy." But such a violent displacement of characters is quite indefensible. Mei Yao-ch'ên is the only commentator who seems to have grasped the meaning: 隨敵一向然後發伏出奇. The *T'u Shu* reads 并力.

we shall succeed in the long run

Literally, "after a thousand *li*."

in killing the commander-in-chief.

Always a great point with the Chinese.

62. This is called ability to accomplish a thing by sheer cunning.

The *T'u Shu* has 是謂巧於成事, and yet another reading,

63. 是故政舉之日夷關折符無通其使
64. 屬於廊廟之上以誅其事

mentioned by Ts'ao Kung, is 巧攻成事. Capt. Calthrop omits
this sentence, after having thus translated the two preceding: "Discover
the enemy's intentions by conforming to his movements. When these
are discovered, then, with one stroke, the general may be killed, even
though he be one hundred leagues distant."

63. On the day that you take up your command,

政舉 does not mean "when war is declared," as Capt. Calthrop
says, nor yet exactly, as Ts'ao Kung paraphrases it, 謀定 "when your
plans are fixed," when you have mapped out your campaign. The phrase
is not given in the *P'ei Wên Yün Fu*. There being no causal connection discoverable between this and the preceding sentence, 是故
must perforce be left untranslated.

block the frontier passes,

夷 is explained by Mei Yao-ch'ên as 滅塞.

destroy the official tallies,

The *locus classicus* for these tallies is *Chou Li*, XIV. fol. 40 (Imperial
edition): 門關用符節貨賄用璽節道路用旌節.
The generic term thus appears to be 節, 符 being the special kind
used at city-gates and on the frontier. They were tablets of bamboo or
wood, one half of which was issued as a permit or passport by the official
in charge of a gate (司門 or 司關. Cf. the 封人 "border-warden"
of *Lun Yü* III. 24, who may have had similar duties.) When this half
was returned to him, within a fixed period, he was authorised to open
the gate and let the traveller through.

and stop the passage of all emissaries.

Either to or from the enemy's country.

64. Be stern in the council-chamber,

Show no weakness, and insist on your plans being ratified by the sovereign.
廊廟 indicates a hall or temple in the Palace. Cf. I. § 26. It is not
clear if other officers would be present. Hardly anything can be made
of 厲, the reading of the standard text, so I have adopted Tu Mu's
conjecture 屬, which appears in the *T'u Shu*.

so that you may control the situation.

65. 敵人開闔必亟入之
66. 先其所愛微與之期

Ts'ao Kung explains 誅 by 治, and Ho Shih by 責成. Another reading is 謀, and Mei Yao-ch'ên, adopting this, understands the whole sentence to mean: Take the strictest precautions to ensure secrecy in your deliberations. Capt. Calthrop glides rather too smoothly over the rough places. His translation is: "conduct the business of the government with vigilance."

65. If the enemy leaves a door open, you must rush in.

This looks a very simple sentence, yet Ts'ao Kung is the only commentator who takes it as I have done. Mêng Shih, followed by Mei Yao-ch'ên and Chang Yü, defines 開闔 as 間者 "spies," and makes 入 an active verb: "If spies come from the enemy, we must quickly let them in." But I cannot find that the words 開闔 have this meaning anywhere else. On the other hand, they may be taken as two verbs, 或開或闔, expressing the enemy's indecision whether to advance or retreat, that being the best moment to attack him. [Cf. *Tao Tê Ching*, chap. X: 天門開闔能爲雌乎; also *Li Chi*, 曲禮, I. ii. 25.] It is not easy to choose between this and Ts'ao Kung's explanation; the fact that 敵人開戶 occurs shortly afterwards, in § 68, might be adduced in support of either. 必 must be understood in the sense of 宜 or 當. The only way to avoid this is to put 開闔 between commas and translate: "If we leave a door open, the enemy is sure to rush in."

66. Forestall your opponent by seizing what he holds dear,

Cf. *supra*, § 18.

and subtly contrive to time his arrival on the ground.

Capt. Calthrop hardly attempts to translate this difficult paragraph, but invents the following instead: "Discover what he most values, and plan to seize it." Ch'ên Hao's explanation, however, is clear enough: 我若先奪便地而敵不至雖有其利亦奚用之是以欲取其愛惜之處必先微與敵人相期誤之使必至 "If I manage to seize a favourable position, but the enemy does not appear on the scene, the advantage thus obtained cannot be turned to any practical account. He who intends, therefore, to occupy a position of importance to the enemy, must begin by making an artful appointment,

67. 踐墨隨敵以決戰事

68. 是故始如處女敵人開戶後如脫兔敵不及拒

so to speak, with his antagonist, and cajole him into going there as well."
Mei Yao-ch'ên explains that this "artful appointment" is to be made
through the medium of the enemy's own spies, who will carry back just
the amount of information that we choose to give them. Then, having
cunningly disclosed our intentions, 我後人發先人至 "we must
manage, though starting after the enemy, to arrive before him" (VII. § 4).
We must start after him in order to ensure his marching thither; we
must arrive before him in order to capture the place without trouble.
Taken thus, the present passage lends some support to Mei Yao-ch'ên's
interpretation of § 47.

67. Walk in the path defined by rule,

墨 stands for 繩墨 "a marking-line," hence a rule of conduct. See
Mencius VII. 1. xli. 2. Ts'ao Kung explains it by the similar metaphor
規矩 "square and compasses." The baldness of the sentiment rather
inclines me to favour the reading 剗 adopted by Chia Lin in place of
踐, which yields an exactly opposite sense, namely: "Discard hard and
fast rules." Chia Lin says: 惟勝是利不可守以繩墨而
為 "Victory is the only thing that matters, and this cannot be achieved
by adhering to conventional canons." It is unfortunate that this variant
rests on very slight authority, for the sense yielded is certainly much
more satisfactory. Napoleon, as we know, according to the veterans of
the old school whom he defeated, won his battles by violating every accepted canon of warfare.

and accommodate yourself to the enemy until you can
fight a decisive battle.

The last four words of the Chinese are omitted by Capt. Calthrop.
Tu Mu says: 隨敵人之形若有可乘之勢則出而
決戰 "Conform to the enemy's tactics until a favourable opportunity
offers; then come forth and engage in a battle that shall prove decisive."

68. At first, then, exhibit the coyness of a maiden,
until the enemy gives you an opening; afterwards emulate
the rapidity of a running hare, and it will be too late
for the enemy to oppose you.

As the hare is noted for its extreme timidity, the comparison hardly appears felicitous. But of course Sun Tzŭ was thinking only of its speed. The words have been taken to mean: You must flee from the enemy as quickly as an escaping hare; but this is rightly rejected by Tu Mu. Capt. Calthrop is wrong in translating 兔 "rabbit." Rabbits are not indigenous to China, and were certainly not known there in the 6th century B.C. The last sixteen characters evidently form a sort of four-line jingle. Chap. X, it may be remembered, closed in similar fashion.

XII. 火攻篇

1. 孫子曰凡火攻有五一曰火人二曰火積三
曰火輜四曰火庫五曰火隊

XII. THE ATTACK BY FIRE.

Rather more than half the chapter (§§ 1—13) is devoted to the subject of fire, after which the author branches off into other topics.

1. Sun Tzŭ said: There are five ways of attacking with fire. The first is to burn soldiers in their camp;

So Tu Mu. Li Ch'üan says: 焚其營殺其士卒也 "Set fire to the camp, and kill the soldiers" (when they try to escape from the flames). Pan Ch'ao, sent on a diplomatic mission to the King of Shan-shan [see XI. § 51, note], found himself placed in extreme peril by the unexpected arrival of an envoy from the Hsiung-nu [the mortal enemies of the Chinese]. In consultation with his officers, he exclaimed: "'Never venture, never win!* The only course open to us now is to make an assault by fire on the barbarians under cover of night, when they will not be able to discern our numbers. Profiting by their panic, we shall exterminate them completely; this will cool the King's courage and cover us with glory, besides ensuring the success of our mission.' The officers all replied that it would be necessary to discuss the matter first with the Intendant (從事). Pan Ch'ao then fell into a passion: 'It is to-day,' he cried, 'that our fortunes must be decided! The Intendant is only a humdrum civilian, who on hearing of our project will certainly be afraid, and everything will be brought to light. An inglorious death is no worthy fate for valiant warriors.' All then agreed to do as he wished. Accordingly, as soon as night came on, he and his little band quickly made their way to the barbarian camp. A strong gale was blowing at the time. Pan Ch'ao ordered ten of the party to take drums and hide behind the enemy's barracks, it being arranged that when they saw flames shoot up, they

---

* 不入虎穴不得虎子 "Unless you enter the tiger's lair, you cannot get hold of the tiger's cubs."

should begin drumming and yelling with all their might. The rest of his men, armed with bows and crossbows, he posted in ambuscade at the gate of the camp. He then set fire to the place from the windward side, whereupon a deafening noise of drums and shouting arose on the front and rear of the Hsiung-nu, who rushed out pell-mell in frantic disorder. Pan Ch'ao slew three of them with his own hand, while his companions cut off the heads of the envoy and thirty of his suite. The remainder, more than a hundred in all, perished in the flames. On the following day, Pan Ch'ao went back and informed 郭恂 Kuo Hsün [the Intendant] of what he had done. The latter was greatly alarmed and turned pale. But Pan Ch'ao, divining his thoughts, said with uplifted hand: 'Although you did not go with us last night, I should not think, Sir, of taking sole credit for our exploit.' This satisfied Kuo Hsün, and Pan Ch'ao, having sent for Kuang, King of Shan-shan, showed him the head of the barbarian envoy. The whole kingdom was seized with fear and trembling, which Pan Ch'ao took steps to allay by issuing a public proclamation. Then, taking the king's son as hostage, he returned to make his report to 竇固 Tou Ku." [*Hou Han Shu*, ch. 47, ff. 1, 2.]

the second is to burn stores;

Tu Mu says: 糧食薪芻 "Provisions, fuel and fodder." In order to subdue the rebellious population of Kiangnan, 高熲 Kao Kêng recommended Wên Ti of the Sui dynasty to make periodical raids and burn their stores of grain, a policy which in the long run proved entirely successful. [隋書, ch. 41, fol. 2.]

the third is to burn baggage-trains;

An example given is the destruction of 袁紹 Yüan Shao's waggons and impedimenta by Ts'ao Ts'ao in 200 A.D.

the fourth is to burn arsenals and magazines;

Tu Mu says that the things contained in 輜 and 庫 are the same. He specifies weapons and other implements, bullion and clothing. Cf. VII. § 11.

the fifth is to hurl dropping fire amongst the enemy.

No fewer than four totally diverse explanations of this sentence are given by the commentators, not one of which is quite satisfactory. It is obvious, at any rate, that the ordinary meaning of 隊 ("regiment" or "company") is here inadmissible. In spite of Tu Mu's note, 焚其行伍因亂而擊之, I must regard "company burning" (Capt. Calthrop's rendering) as nonsense pure and simple. We may also, I think, reject the very forced explanation given by Li Ch'üan, Mei Yao-ch'ên

2. 行火必有因煙火必素具

3. 發火有時起火有日

and Chang Yü, of whom the last-named says: 焚其隊仗使兵無戰具 "burning a regiment's weapons, so that the soldiers may have nothing to fight with." That leaves only two solutions open: one, favoured by Chia Lin and Ho Shih, is to take 隊 in the somewhat uncommon sense of "a road," = 隧. The commentary on a passage in the 穆天子傳, quoted in *K'ang Hsi*, defines 隊 (read *sui*) as 谷中險阻道 "a difficult road leading through a valley." Here it would stand for the 糧道 "line of supplies," which might be effectually interrupted if the country roundabout was laid waste with fire. Finally, the interpretation which I have adopted is that given by Tu Yu in the *T'ung Tien*. He reads 墜 (which is not absolutely necessary, 隊 *chui* being sometimes used in the same sense), with the following note: 以火墮敵營中也火墜之法以鐵籠火着箭頭頸強弩射敵營中 "To drop fire into the enemy's camp. The method by which this may be done is to set the tips of arrows alight by dipping them into a brazier, and then shoot them from powerful crossbows into the enemy's lines."

2. In order to carry out an attack with fire, we must have means available.

Ts'ao Kung thinks that 姦人 "traitors in the enemy's camp" are referred to. He thus takes 因 as the efficient cause only. But Ch'ên Hao is more likely to be right in saying: 須得其便不獨姦人 "We must have favourable circumstances in general, not merely traitors to help us." Chia Lin says: 因風燥 "We must avail ourselves of wind and dry weather."

the material for raising fire should always be kept in readiness.

煙火 is explained by Ts'ao Kung as 燒具 "appliances for making fire." Tu Mu suggests 艾蒿荻葦薪芻膏油之屬 "dry vegetable matter, reeds, brushwood, straw, grease, oil, etc." Here we have the material cause. Chang Yü says: 貯火之器燃火之物 "vessels for hoarding fire, stuff for lighting fires."

3. There is a proper season for making attacks with fire, and special days for starting a conflagration.

4. 時者天之燥也日者宿在箕壁翼軫也凡此
　四宿者風起之日也

5. 凡火攻必因五火之變而應之

6. 火發於內則早應之於外

A fire must not be begun 妄 "recklessly" or 偶然 "at haphazard."

4. The proper season is when the weather is very dry;
the special days are those when the moon is in the constellations of the Sieve, the Wall, the Wing or the Cross-bar;

These are, respectively, the 7th, 14th, 27th, and 28th of the 二十八宮 Twenty-eight Stellar Mansions, corresponding roughly to Sagittarius, Pegasus, Crater and Corvus. The original text, followed by the *T'u Shu*, has 月 in place of 宿; the present reading rests on the authority of the *T'ung Tien* and *Yü Lan*. Tu Mu says: 宿者月之所宿也. For 箕壁, both *T'ung Tien* and *Yü Lan* give the more precise location 戊箕東壁. Mei Yao-ch'ên tells us that by 箕 is meant the tail of the 龍 Dragon; by 壁, the eastern part of that constellation; by 翼 and 軫, the tail of the 鶉 Quail.

for these four are all days of rising wind.

此四宿者 is elliptical for 月在此四宿之日. 蕭繹 Hsiao I (afterwards fourth Emperor of the Liang dynasty, A.D. 552—555) is quoted by Tu Yu as saying that the days 丙丁 of spring, 戊巳 of summer, 壬癸 of autumn, and 甲乙 of winter bring fierce gales of wind and rain.

5. In attacking with fire, one should be prepared to meet five possible developments:

I take 五 as qualifying 變, not 火, and therefore think that Chang Yü is wrong in referring 五火 to the five methods of attack set forth in § 1. What follows has certainly nothing to do with these.

6. (1) When fire breaks out inside the enemy's camp, respond at once

The *Yü Lan* incorrectly reads 軍 for 早.

with an attack from without.

7. 火發而其兵靜者待而勿攻
8. 極其火力可從而從之不可從而止
9. 火可發於外無待於內以時發之

7. (2) If there is an outbreak of fire, but the enemy's soldiers remain quiet, bide your time and do not attack.

The original text omits 而其. The prime object of attacking with fire is to throw the enemy into confusion. If this effect is not produced, it means that the enemy is ready to receive us. Hence the necessity for caution.

8. (3) When the force of the flames has reached its height, follow it up with an attack, if that is practicable; if not, stay where you are.

Ts'ao Kung says: 見可而進知難而退 "If you see a possible way, advance; but if you find the difficulties too great, retire."

9. (4) If it is possible to make an assault with fire from without, do not wait for it to break out within, but deliver your attack at a favourable moment.

Tu Mu says that the previous paragraphs had reference to the fire breaking out (either accidentally, we may suppose, or by the agency of incendiaries) inside the enemy's camp. "But," he continues, 若敵居荒澤草穢或營柵可焚之地卽須及時發火不必更待內發作然後應之恐敵人自燒野草我起火無益 "if the enemy is settled in a waste place littered with quantities of grass, or if he has pitched his camp in a position which can be burnt out, we must carry our fire against him at any seasonable opportunity, and not wait on in hopes of an outbreak occurring within, for fear our opponents should themselves burn up the surrounding vegetation, and thus render our own attempts fruitless." The famous 李陵 Li Ling once baffled the 單于 leader of the Hsiung-nu in this way. The latter, taking advantage of a favourable wind, tried to set fire to the Chinese general's camp, but found that every scrap of combustible vegetation in the neighbourhood had already been burnt down. On the other hand, 波才 Po-ts'ai, a general of the 黃巾賊 Yellow Turban rebels, was badly defeated in 184 A.D. through his neglect of this simple precaution. "At the head of a large army he was besieging 長社 Ch'ang-shê, which was held by 皇甫嵩 Huang-fu Sung. The garrison was very

10. 火發上風無攻下風
11. 晝風久夜風止
12. 凡軍必知有五火之變以數守之

small, and a general feeling of nervousness pervaded the ranks; so Huang-fu Sung called his officers together and said: 'In war, there are various indirect methods of attack, and numbers do not count for everything. [The commentator here quotes Sun Tzŭ, V. §§ 5, 6 and 10.] Now the rebels have pitched their camp in the midst of thick grass (依草結營), which will easily burn when the wind blows. If we set fire to it at night, they will be thrown into a panic, and we can make a sortie and attack them on all sides at once, thus emulating the achievement of T'ien Tan.' [See p. 90.] That same evening, a strong breeze sprang up; so Huang-fu Sung instructed his soldiers to bind reeds together into torches and mount guard on the city walls, after which he sent out a band of daring men, who stealthily made their way through the lines and started the fire with loud shouts and yells. Simultaneously, a glare of light shot up from the city-walls, and Huang-fu Sung, sounding his drums, led a rapid charge, which threw the rebels into confusion and put them to headlong flight." [*Hou Han Shu*, ch. 71, f. 2 r°.]

10. (5) When you start a fire, be to windward of it. Do not attack from the leeward.

Chang Yü, following Tu Yu, says: 燒之必退退而逆擊之必死戰則不便也 "When you make a fire, the enemy will retreat away from it; if you oppose his retreat and attack him then, he will fight desperately, which will not conduce to your success." A rather more obvious explanation is given by Tu Mu: "If the wind is in the east, begin burning to the east of the enemy, and follow up the attack yourself from that side. If you start the fire on the east side, and then attack from the west, you will suffer in the same way as your enemy."

11. A wind that rises in the daytime lasts long, but a night breeze soon falls.

Cf. Lao Tzŭ's saying: 飄風不終朝 "A violent wind does not last the space of a morning." (*Tao Tê Ching*, chap. 23.) Mei Yao-ch'ên and Wang Hsi say: "A day breeze dies down at nightfall, and a night breeze at daybreak. This is what happens as a general rule." The phenomenon observed may be correct enough, but how this sense is to be obtained is not apparent.

12. In every army, the five developments connected with fire must be known, the movements of the stars calculated, and a watch kept for the proper days.

13. 故以火佐攻者明以水佐攻者強

14. 水可以絶不可以奪

Tu Mu's commentary shows what has to be supplied in order to make sense out of 以數守之. He says: 須籌星躔之數守風起之日乃可發火 "We must make calculations as to the paths of the stars, and watch for the days on which wind will rise, before making our attack with fire." Chang Yü seems to take 守 in the sense of 防: "We must not only know how to assail our opponents with fire, but also be on our guard against similar attacks from them."

13. Hence those who use fire as an aid to the attack show intelligence;

I have not the least hesitation in rejecting the commentators' explanation of 明 as = 明白. Thus Chang Yü says: 灼然可以取勝 "...will *clearly* [i.e. obviously] be able to gain the victory." This is not only clumsy in itself, but does not balance 強 in the next clause. For 明 "intelligent," cf. *infra*, § 16, and *Lun Yü* XII. 6.

those who use water as an aid to the attack gain an accession of strength.

Capt. Calthrop gives an extraordinary rendering of the paragraph: "...if the attack is to be assisted, the fire must be unquenchable. If water is to assist the attack, the flood must be overwhelming."

14. By means of water, an enemy may be intercepted, but not robbed of all his belongings.

Ts'ao Kung's note is: 但可以絶敵道分敵軍不可以奪敵蓄積 "We can merely obstruct the enemy's road or divide his army, but not sweep away all his accumulated stores." Water can do useful service, but it lacks the terrible destructive power of fire. This is the reason, Chang Yü concludes, why the former is dismissed in a couple of sentences, whereas the attack by fire is discussed in detail. Wu Tzŭ (ch. 4) speaks thus of the two elements: 居軍下濕水無所通霖雨數至可灌而沉居軍荒澤草楚幽穢風飆數至可焚而滅 "If an army is encamped on low-lying marshy ground, from which the water cannot run off, and where the rainfall is heavy, it may be submerged by a flood. If an army is encamped in wild marsh lands thickly overgrown with weeds and brambles, and visited by frequent gales, it may be exterminated by fire."

15. 夫戰勝攻取而不修其功者凶命曰費留
16. 故曰明主慮之良將修之

15. Unhappy is the fate of one who tries to win his battles and succeed in his attacks without cultivating the spirit of enterprise; for the result is waste of time and general stagnation.

This is one of the most perplexing passages in Sun Tzŭ. The difficulty lies mainly in 不修其功, of which two interpretations appear possible. Most of the commentators understand 修 in the sense (not known to *K'ang Hsi*) of 賞 "reward" or 舉 "promote," and 其功 as referring to the merit of officers and men. Thus Ts'ao Kung says: 賞善不踰日 "Rewards for good service should not be deferred a single day." And Tu Mu: "If you do not take opportunity to advance and reward the deserving, your subordinates will not carry out your commands, and disaster will ensue." 費留 would then probably mean 留滯費耗 "stoppage of expenditure," or as Chia Lin puts it, 惜費 "the grudging of expenditure." For several reasons, however, and in spite of the formidable array of scholars on the other side, I prefer the interpretation suggested by Mei Yao-ch'ên alone, whose words I will quote: 欲戰必勝攻必取者在因時乘便能作爲功也 作爲功者修火攻水攻之類不可坐守其利也 坐守其利者凶也 "Those who want to make sure of succeeding in their battles and assaults must seize the favourable moments when they come and not shrink on occasion from heroic measures: that is to say, they must resort to such means of attack as fire, water and the like. What they must not do, and what will prove fatal, is to sit still and simply hold on to the advantages they have got." This retains the more usual meaning of 修, and also brings out a clear connection of thought with the previous part of the chapter. With regard to 費留, Wang Hsi paraphrases it as 費財老師 "expending treasure and tiring out [*lit.*, ageing] the army." 費 of course is expenditure or waste in general, either of time, money or strength. But the soldier is less concerned with the saving of money than of time. For the metaphor expressed in "stagnation" I am indebted to Ts'ao Kung, who says: 若水之留不復還也. Capt. Calthrop gives a rendering which bears but little relation to the Chinese text: "unless victory or possession be obtained, the enemy quickly recovers, and misfortunes arise. The war drags on, and money is spent."

16. Hence the saying: The enlightened ruler lays his plans well ahead; the good general cultivates his resources.

17. 非利不動非得不用非危不戰

18. 主不可以怒而興師將不可以慍而致戰

19. 合於利而動不合於利而止

As Sun Tzŭ quotes this jingle in support of his assertion in § 15, we must suppose 修之 to stand for 修其功 or something analogous. The meaning seems to be that the ruler lays plans which the general must show resourcefulness in carrying out. It is now plainer than ever that 修 cannot mean "to reward." Nevertheless, Tu Mu quotes the following from the 三略, ch. 2: 霸者制士以權結士以 信使士以賞信衰則士疏賞虧則士不用命 "The warlike prince controls his soldiers by his authority, knits them together by good faith, and by rewards makes them serviceable. If faith decays, there will be disruption; if rewards are deficient, commands will not be respected."

17. Move not unless you see an advantage;

起, the *Yü Lan's* variant for 動, is adopted by Li Ch'üan and Tu Mu.

use not your troops unless there is something to be gained; fight not unless the position is critical.

Sun Tzŭ may at times appear to be over-cautious, but he never goes so far in that direction as the remarkable passage in the *Tao Té Ching*, ch. 69: 吾不敢為主而為客不敢進寸而退尺 "I dare not take the initiative, but prefer to act on the defensive; I dare not advance an inch, but prefer to retreat a foot."

18. No ruler should put troops into the field merely to gratify his own spleen; no general should fight a battle simply out of pique.

Again compare Lao Tzŭ, ch. 68: 善戰者不怒. Chang Yü says that 慍 is a weaker word than 怒, and is therefore applied to the general as opposed to the sovereign. The *T'ung Tien* and *Yü Lan* read 軍 for 師, and the latter 合 for 致.

19. If it is to your advantage, make a forward move; if not, stay where you are.

This is repeated from XI. § 17. Here I feel convinced that it is an interpolation, for it is evident that § 20 ought to follow immediately on

20. 怒可以復喜慍可以復悅
21. 亡國不可以復存死者不可以復生
22. 故明君慎之良將警之此安國全軍之道也

§ 18. For 動, the *T'ung Tien* and *Yü Lan* have 用. Capt. Calthrop invents a sentence which he inserts before this one: "Do not make war unless victory may be gained thereby." While he was about it, he might have credited Sun Tzŭ with something slightly less inane.

20. Anger may in time change to gladness; vexation may be succeeded by content.

According to Chang Yü, 喜 denotes joy outwardly manifested in the countenance, 悅 the inward sensation of happiness.

21. But a kingdom that has once been destroyed can never come again into being;

The Wu State was destined to be a melancholy example of this saying. See p. 50.

nor can the dead ever be brought back to life.

22. Hence the enlightened ruler is heedful, and the good general full of caution.

警, which usually means "to warn," is here equal to 戒. This is a good instance of how Chinese characters, which stand for ideas, refuse to be fettered by dictionary-made definitions. The *T'u Shu* reads 故曰, as in § 16.

This is the way to keep a country at peace and an army intact.

It is odd that 全軍 should not have the same meaning here as in III. § 1, *q. v.* This has led me to consider whether it might not be possible to take the earlier passage thus: "to preserve your own army (country, regiment, etc.) intact is better than to destroy the enemy's." The two words do not appear in the *T'ung Tien* or the *Yü Lan*. Capt. Calthrop misses the point by translating: "then is the state secure, and the army victorious in battle."

XIII. 用間篇

1. 孫子曰凡興師十萬出兵千里百姓之費公
家之奉日費千金內外騷動怠於道路不得
操事者七十萬家

XIII. THE USE OF SPIES.

間 is really a vulgar form of 閒, and does not appear in the *Shuo Wĕn*. In practice, however, it has gradually become a distinct character with special meanings of its own, and I have therefore followed my edition of the standard text in retaining this form throughout the chapter. In VI. § 25, on the other hand, the correct form 閒 will be found. The evolution of the meaning "spy" is worth considering for a moment, provided it be understood that this is very doubtful ground, and that any dogmatism is out of place. The *Shuo Wĕn* defines 閒 as 隟 (the old form of 隙) "a crack" or "chink," and on the whole we may accept 徐鍇 Hsü Ch'ieh's analysis as not unduly fanciful: 夫門夜閉 閉而見月光是有閒隟也 "At night, a *door* is shut; if, when it is shut, the light of the *moon* is visible, it must come through a *chink*." From this it is an easy step to the meaning "space between," or simply "between," as for example in the phrase 往來閒諜 "to act as a secret spy between enemies." Here 諜 is the word which means "spy;" but we may suppose that constant association so affected the original force of 閒, that 諜 could at last be dropped altogether, leaving 閒 to stand alone with the same signification. Another possible theory is that the word may first have come to mean 覷 "to peep" (see 博雅, quoted in *K'ang Hsi*), which would naturally be suggested by "crack" or "crevice," and afterwards the man who peeps, or spy.

1. Sun Tzŭ said: Raising a host of a hundred thousand men and marching them great distances entails heavy loss on the people and a drain on the resources of the State. The daily expenditure will amount to a thousand ounces of silver.

2. 相守數年以爭一日之勝而愛爵祿百金不知敵之情者不仁之至也

Cf. II. §§ 1, 13, 14.

There will be commotion at home and abroad, and men will drop down exhausted on the highways.

怠 於 道 路, which is omitted by the *Yü Lan*, appears at first sight to be explained by the words immediately following, so that the obvious translation would be "(enforced) idleness along the line of march." [Cf. *Tao Tê Ching*, ch. 30: 師 之 所 處 荊 棘 生 焉 "Where troops have been quartered, brambles and thorns spring up."] The commentators, however, say that 怠 is here equivalent to 疲 — a meaning which is still retained in the phrase 倦 怠. Tu Mu refers 怠 to those who are engaged in conveying provisions to the army. But this can hardly be said to emerge clearly from Sun Tzŭ's text. Chang Yü has the note: "We may be reminded of the saying: 'On serious ground, gather in plunder' [XI. § 13]. Why then should carriage and transportation cause exhaustion on the highways? — The answer is, that not victuals alone, but all sorts of munitions of war have to be conveyed to the army. Besides, the injunction to 'forage on the enemy' only means that when an army is deeply engaged in hostile territory, scarcity of food must be provided against. Hence, without being solely dependent on the enemy for corn, we must forage in order that there may be an uninterrupted flow of supplies. Then, again, there are places like salt deserts (磧 鹵 之 地), where provisions being unobtainable, supplies from home cannot be dispensed with."

As many as seven hundred thousand families will be impeded in their labour.

Mei Yao-ch'ên says: 廢 於 耒 耜 "Men will be lacking at the plough-tail." The allusion is to 井 田 the system of dividing land into nine parts, as shown in the character 井, each consisting of a 夫 or 頃 (about 15 acres), the plot in the centre being cultivated on behalf of the State by the tenants of the other eight. It was here also, so Tu Mu tells us, that their cottages were built and a well sunk, to be used by all in common. [See II. § 12, note.] These groups of eight peasant proprietors were called 鄰. In time of war, one of the families had to serve in the army, while the other seven contributed to its support (一 家 從 軍 七 家 奉 之). Thus, by a levy of 100,000 men (reckoning one able-bodied soldier to each family) the husbandry of 700,000 families would be affected.

2. Hostile armies may face each other for years, striving

3. 非人之將也非主之佐也非勝之主也

for the victory which is decided in a single day. This being so, to remain in ignorance of the enemy's condition simply because one grudges the outlay of a hundred ounces of silver in honours and emoluments,

"For spies" is of course the meaning, though it would spoil the effect of this curiously elaborate exordium if spies were actually mentioned at this point.

is the height of inhumanity.

Sun Tzŭ's argument is certainly ingenious. He begins by adverting to the frightful misery and vast expenditure of blood and treasure which war always brings in its train. Now, unless you are kept informed of the enemy's condition, and are ready to strike at the right moment, a war may drag on for years. The only way to get this information is to employ spies, and it is impossible to obtain trustworthy spies unless they are properly paid for their services. But it is surely false economy to grudge a comparatively trifling amount for this purpose, when every day that the war lasts eats up an incalculably greater sum. This grievous burden falls on the shoulders of the poor, and hence Sun Tzŭ concludes that to neglect the use of spies is nothing less than a crime against humanity.

3. One who acts thus is no leader of men, no present help to his sovereign,

An inferior reading for 主 is 仁, thus explained by Mei Yao-ch'ên:

非以仁佐國者也.

no master of victory.

This idea, that the true object of war is peace, has its root in the national temperament of the Chinese. Even so far back as 597 B.C., these memorable words were uttered by Prince 莊 Chuang of the Ch'u State:

夫文止戈爲武…夫武禁暴戢兵保大定功安民和衆豐財者也 "The character for 'prowess' (武) is made up of 止 'to stay' and 戈 'a spear' (cessation of hostilities). Military prowess is seen in the repression of cruelty, the calling in of weapons, the preservation of the appointment of Heaven, the firm establishment of merit, the bestowal of happiness on the people, putting harmony between the princes, the diffusion of wealth." [*Tso Chuan*, 宣公 XII. 3 *ad fin.*]

4. 故明君賢將所以動而勝人成功出於眾者
先知也

5. 先知者不可取於鬼神不可象於事不可驗
於度

6. 必取於人知敵之情者也

4. Thus, what enables the wise sovereign and the good
general to strike and conquer, and achieve things beyond
the reach of ordinary men, is *foreknowledge*.

That is, knowledge of the enemy's dispositions, and what he means to do.

5. Now this foreknowledge cannot be elicited from spirits;

以禱祀 "by prayers or sacrifices," says Chang Yü. 鬼 are the
disembodied spirits of men, and 神 supernatural beings or "gods."

it cannot be obtained inductively from experience,

Tu Mu's note makes the meaning clear: 象, he says, is the same as
類 reasoning by analogy; 不可以他事比類而求 "[know-
ledge of the enemy] cannot be gained by reasoning from other analogous cases."

nor by any deductive calculation.

Li Ch'üan says: 夫長短闊狹遠近小大即可驗之
於度數人之情偽度不能知也 "Quantities like length,
breadth, distance and magnitude, are susceptible of exact mathematical
determination; human actions cannot be so calculated."

6. Knowledge of the enemy's dispositions can only be
obtained from other men.

Mei Yao-ch'ên has rather an interesting note: 鬼神之情可以
筮卜知形氣之物可以象類求天地之理可以
度數驗唯敵之情必由間者而後知也 "Know-
ledge of the spirit-world is to be obtained by divination; information in
natural science may be sought by inductive reasoning; the laws of the
universe can be verified by mathematical calculation: but the dispositions
of an enemy are ascertainable through spies and spies alone."

7. 故用間有五有鄉間有內間有反間有死間
有生間

8. 五間俱起莫知其道是謂神紀人君之寶也

9. 鄉間者因其鄉人而用之

7. Hence the use of spies, of whom there are five classes: (1) Local spies; (2) inward spies; (3) converted spies; (4) doomed spies; (5) surviving spies.

8. When these five kinds of spy are all at work, none can discover the secret system.

道 is explained by Tu Mu as 其情泄形露之道 "the way in which facts leak out and dispositions are revealed."

This is called

為 is the reading of the standard text, but the *T'ung Tien, Yü Lan* and *T'u Shu* all have 謂.

"divine manipulation of the threads."

Capt. Calthrop translates 神紀 "the Mysterious Thread," but Mei Yao-ch'ên's paraphrase 神妙之綱紀 shows that what is meant is the *control* of a number of threads.

It is the sovereign's most precious faculty.

"Cromwell, one of the greatest and most practical of all cavalry leaders, had officers styled 'scout masters,' whose business it was to collect all possible information regarding the enemy, through scouts and spies, etc., and much of his success in war was traceable to the previous knowledge of the enemy's moves thus gained." *

9. Having *local spies*

鄉間 is the emended reading of Chia Lin and the *T'u Shu* for the unintelligible 因間, here and in § 7, of the standard text, which nevertheless reads 鄉間 in § 22.

means employing the services of the inhabitants of a district.

Tu Mu says: "In the enemy's country, win people over by kind treatment, and use them as spies."

---

* "Aids to Scouting," p. 2.

10. 內間者因其官人而用之

10. Having *inward spies*, making use of officials of the enemy.

官 includes both civil and military officials. Tu Mu enumerates the following classes as likely to do good service in this respect: "Worthy men who have been degraded from office, criminals who have undergone punishment; also, favourite concubines who are greedy for gold, men who are aggrieved at being in subordinate positions, or who have been passed over in the distribution of posts, others who are anxious that their side should be defeated in order that they may have a chance of displaying their ability and talents, fickle turncoats who always want to have a foot in each boat (覆覆變詐常持兩端之心者). Officials of these several kinds," he continues, "should be secretly approached and bound to one's interests by means of rich presents. In this way you will be able to find out the state of affairs in the enemy's country, ascertain the plans that are being formed against you, and moreover disturb the harmony and create a breach between the sovereign and his ministers." The necessity for extreme caution, however, in dealing with "inward spies," appears from an historical incident related by Ho Shih: "羅尚 Lo Shang, Governor of 益州 I-chou, sent his general 隗伯 Wei Po to attack the rebel 李雄 Li Hsiung of 蜀 Shu in his stronghold at 郫 P'i. After each side had experienced a number of victories and defeats, Li Hsiung had recourse to the services of a certain 朴泰 P'o-t'ai, a native of 武都 Wu-tu. He began by having him whipped until the blood came, and then sent him off to Lo Shang, whom he was to delude by offering to co-operate with him from inside the city, and to give a fire signal at the right moment for making a general assault. Lo Shang, confiding in these promises, marched out all his best troops, and placed Wei Po and others at their head with orders to attack at P'o-t'ai's bidding. Meanwhile, Li Hsiung's general, 李驤 Li Hsiang, had prepared an ambuscade on their line of march; and P'o-t'ai, having reared long scaling-ladders against the city walls, now lighted the beacon-fire. Wei Po's men raced up on seeing the signal and began climbing the ladders as fast as they could, while others were drawn up by ropes lowered from above. More than a hundred of Lo Shang's soldiers entered the city in this way, every one of whom was forthwith beheaded. Li Hsiung then charged with all his forces, both inside and outside the city, and routed the enemy completely." [This happened in 303 A.D. I do not know where Ho Shih got the story from. It is not given in the biography of Li Hsiung or that of his father Li 特 T'ê, *Chin Shu*, ch. 120, 121.]

11. 反間者因其敵間而用之

11. Having *converted spies*, getting hold of the enemy's spies and using them for our own purposes.

By means of heavy bribes and liberal promises detaching them from the enemy's service, and inducing them to carry back false information as well as to spy in turn on their own countrymen. Thus Tu Yu: 因厚賂重許反使爲我間也. On the other hand, 蕭世誠 Hsiao Shih-hsien in defining the 反間 says that we pretend not to have detected him, but contrive to let him carry away a false impression of what is going on (敵使人來候我我佯不知而示以虛事). Several of the commentators accept this as an alternative definition; but that it is not what Sun Tzŭ meant is conclusively proved by his subsequent remarks about treating the converted spy generously (§ 21 *sqq.*). Ho Shih notes three occasions on which converted spies were used with conspicuous success: 1) by T'ien Tan in his defence of Chi-mo (see *supra*, p. 90); 2) by Chao Shê on his march to O-yü (see p. 57); and by the wily 范雎 Fan Chü in 260 B.C., when Lien P'o was conducting a defensive campaign against Ch'in. The King of Chao strongly disapproved of Lien P'o's cautious and dilatory methods, which had been unable to avert a series of minor disasters, and therefore lent a ready ear to the reports of his spies, who had secretly gone over to the enemy and were already in Fan Chü's pay. They said: "The only thing which causes Ch'in anxiety is lest 趙括 Chao Kua should be made general. Lien P'o they consider an easy opponent, who is sure to be vanquished in the long run." Now this Chao Kua was a son of the famous Chao Shê. From his boyhood, he had been wholly engrossed in the study of war and military matters, until at last he came to believe that there was no commander in the whole Empire who could stand against him. His father was much disquieted by this overweening conceit, and the flippancy with which he spoke of such a serious thing as war, and solemnly declared that if ever Kua was appointed general, he would bring ruin on the armies of Chao. This was the man who, in spite of earnest protests from his own mother and the veteran statesman 藺相如 Lin Hsiang-ju, was now sent to succeed Lien P'o. Needless to say, he proved no match for the redoubtable Po Ch'i and the great military power of Ch'in. He fell into a trap by which his army was divided into two and his communications cut; and after a desperate resistance lasting 46 days, during which the famished soldiers devoured one another, he was himself killed by an arrow, and his whole force, amounting, it is said, to 400,000 men, ruthlessiy put to the sword. [See 歷代紀事年表, ch. 19, ff. 48—50].

12. 死間者爲誑事於外令吾間知之而傳於敵
13. 生間者反報也

12. Having *doomed spies*, doing certain things openly for purposes of deception, and allowing our own spies to know of them and report them to the enemy.

傳 is Li Ch'üan's conjecture for 待, which is found in the *T'ung Tien* and the *Yü Lan*. The *Tu Shu*, unsupported by any good authority, adds 間也 after 敵. In that case, the doomed spies would be those of the enemy, to whom our own spies had conveyed false information. But this is unnecessarily complicated. Tu Yu gives the best exposition of the meaning: "We ostentatiously do things calculated to deceive our own spies, who must be led to believe that they have been unwittingly disclosed. Then, when these spies are captured in the enemy's lines, they will make an entirely false report, and the enemy will take measures accordingly, only to find that we do something quite different. The spies will thereupon be put to death." Capt. Calthrop makes a hopeless muddle of the sentence. As an example of doomed spies, Ho Shih mentions the prisoners released by Pan Ch'ao in his campaign against Yarkand. (See p. 132.) He also refers to 唐儉 T'ang Chien, who in 630 A.D. was sent by T'ai Tsung to lull the Turkish Khan 頡利 Chieh-li into fancied security, until Li Ching was able to deliver a crushing blow against him. Chang Yü says that the Turks revenged themselves by killing T'ang Chien, but this is a mistake, for we read in both the Old and the New T'ang History (ch. 58, fol. 2 and ch. 89, fol. 8 respectively) that he escaped and lived on until 656. 酈食其 Li I-chi* played a somewhat similar part in 203 B.C., when sent by the King of Han to open peaceful negotiations with Ch'i. He has certainly more claim to be described as a 死間; for the King of Ch'i, being subsequently attacked without warning by Han Hsin, and infuriated by what he considered the treachery of Li I-chi, ordered the unfortunate envoy to be boiled alive.

13. *Surviving spies*, finally, are those who bring back news from the enemy's camp.

This is the ordinary class of spies, properly so called, forming a regular part of the army. Tu Mu says: 生間者必取內明外愚形劣心壯趫健勁勇閑於鄙事能忍饑寒垢耻者爲之 "Your surviving spy must be a man of keen intellect, though

---

* *Ch'ien Han Shu*, ch. 43, fol. 1. 顏師古 Yen Shih-ku *in loc*. says: 食音異其音基.

14. 故三軍之親莫親於間賞莫厚於間事莫密於間

in outward appearance a fool; of shabby exterior, but with a will of iron. He must be active, robust, endowed with physical strength and courage; thoroughly accustomed to all sorts of dirty work, able to endure hunger and cold, and to put up with shame and ignominy." Ho Shih tells the following story of 達奚武 Ta-hsi Wu of the Sui dynasty: "When he was governor of Eastern Ch'in, 神武 Shên-wu of Ch'i made a hostile movement upon 沙苑 Sha-yüan. The Emperor T'ai Tsu [? Kao Tsu] sent Ta-hsi Wu to spy upon the enemy. He was accompanied by two other men. All three were on horseback and wore the enemy's uniform. When it was dark, they dismounted a few hundred feet away from the enemy's camp and stealthily crept up to listen, until they succeeded in catching the passwords used by the army. Then they got on their horses again and boldly passed through the camp under the guise of night-watchmen (警夜者); and more than once, happening to come across a soldier who was committing some breach of discipline, they actually stopped to give the culprit a sound cudgelling! Thus they managed to return with the fullest possible information about the enemy's dispositions, and received warm commendation from the Emperor, who in consequence of their report was able to inflict a severe defeat on his adversary." With the above classification it is interesting to compare the remarks of Frederick the Great: * "Es giebt vielerley Sorten von Spions: 1. Geringe Leute, welche sich von diesem Handwerk meliren. 2. Doppelte Spions. 3. Spions von Consequenz, und endlich 4. Diejenigen, welche man zu diesem unglücklichen Hankwerk zwinget." This of course is a bad cross-division. The first class ("Bürgersleute, Bauern, Priesters, etc.") corresponds roughly to Sun Tzŭ's "local spies," and the third to "inward spies." Of "Doppelte Spions" it is broadly stated that they are employed "um dem Feinde falsche Nachrichten aufzubinden." Thus they would include both converted and doomed spies. Frederick's last class of spies does not appear in Sun Tzŭ's list, perhaps because the risk in using them is too great.

14. Hence it is that with none in the whole army are more intimate relations to be maintained than with spies.

The original text and the *T'u Shu* have 事 in place of the first 親. Tu Mu and Mei Yao-ch'ên point out that the spy is privileged to enter even the general's private sleeping-tent. Capt. Calthrop has an inaccurate translation: "In connection with the armies, spies should be treated with the greatest kindness."

---

* "Unterricht des Königs von Preussen an die Generale seiner Armeen," cap. 12 (edition of 1794).

15. 非聖智不能用間

None should be more liberally rewarded.

Frederick concludes his chapter on spies with the words: "Zu allem diesem füge ich noch hinzu, dass man in Bezahlung der Spions freygebig, ja verschwenderisch seyn muss. Ein Mench, der um eures Dienstes halber den Strick waget, verdienet dafür belohnet zu werden."

In no other business should greater secrecy be preserved.

Tu Mu gives a graphic touch: 出口入耳也, that is to say, all communications with spies should be carried on "mouth-to-ear." Capt. Calthrop has: "All matters relating to spies are secret," which is distinctly feeble. An inferior reading for 密 is 審. The following remarks on spies may be quoted from Turenne, who made perhaps larger use of them than any previous commander: "Spies are attached to those who give them most, he who pays them ill is never served. They should never be known to anybody; nor should they know one another. When they propose anything very material, secure their persons, or have in your possession their wives and children as hostages for their fidelity. Never communicate anything to them but what it is absolutely necessary that they should know." *

15. Spies cannot be usefully employed

This is the *nuance* of Tu Yu's paraphrase 不能得間人之用.

without a certain intuitive sagacity.

Mei Yao-ch'ên says: 知其情僞辨其邪正則能用 "In order to use them, one must know fact from falsehood, and be able to discriminate between honesty and double-dealing." Wang Hsi takes 聖 and 智 separately, defining the former as 通而先識 "intuitive perception" and the latter as 明於事 "practical intelligence." Tu Mu strangely refers these attributes to the spies themselves: 先量間者之性誠實多智然後可用之 "Before using spies we must assure ourselves as to their integrity of character and the extent of their experience and skill." But he continues: 厚貌深情險於山川非聖人莫能知 "A brazen face and a crafty disposition are more dangerous than mountains or rivers; it takes a man of genius to penetrate such." So that we are left in some doubt as to his real opinion on the passage.

16. 非仁義不能使間
17. 非微妙不能得間之實
18. 微哉微哉無所不用間也
19. 間事未發而先聞者間與所告者皆死

16. They cannot be properly managed without benevolence and straightforwardness.

Chang Yü says that 仁 means "not grudging them honours and pay;"
義, "showing no distrust of their honesty." "When you have attracted
them by substantial offers, you must treat them with absolute sincerity;
then they will work for you with all their might."

17. Without subtle ingenuity of mind, one cannot make
certain of the truth of their reports.

Mei Yao-ch'ên says: "Be on your guard against the possibility of spies
going over to the service of the enemy." The *T'ung Tien* and *Yü Lan*
read 密 for 妙.

18. Be subtle! be subtle!

Cf. VI. § 9: 微乎微乎. Capt. Calthrop translates: "Wonderful
indeed is the power of spies."

and use your spies for every kind of business.

19. If a secret piece of news is divulged by a spy
before the time is ripe, he must be put to death together
with the man to whom the secret was told.

The Chinese here is so concise and elliptical that some expansion is
necessary for the proper understanding of it. 間事 denotes important
information about the enemy obtained from a surviving spy. The subject of 未發, however, is not this information itself, but the secret
stratagem built up on the strength of it. 聞者 means "is heard" —
by anybody else. Thus, word for word, we get: "If spy matters are
heard before [our plans] are carried out," etc. Capt. Calthrop, in translating 間與所告者 "the spy who told the matter, and the man
who repeated the same," may appeal to the authority of the commentators; but he surely misses the main point of Sun Tzŭ's injunction.
For, whereas you kill the spy himself 惡其泄 "as a punishment for
letting out the secret," the object of killing the other man is only, as
Ch'ên Hao puts it, 以滅口 "to stop his mouth" and prevent the

20. 凡軍之所欲擊城之所欲攻人之所欲殺必
先知其守將左右謁者門者舍人之姓名令
吾間必索知之

news leaking any further. If it had already been repeated to others, this object would not be gained. Either way, Sun Tzǔ lays himself open to the charge of inhumanity, though Tu Mu tries to defend him by saying that the man deserves to be put to death, for the spy would certainly not have told the secret unless the other had been at pains to worm it out of him. The *T'ung Tien* and *Yü Lan* have the reading . . . 先聞其間者與, etc., which, while not affecting the sense, strikes me as being better than that of the standard text. The *T'u Shu* has . . . 聞與所告者, which I suppose would mean: "the man who heard the secret and the man who told it to him."

20. Whether the object be to crush an army, to storm a city, or to assassinate an individual, it is always necessary to begin by finding out the names of the attendants,

左右 is a comprehensive term for those who wait on others, servants and retainers generally. Capt. Calthrop is hardly happy in rendering it "right-hand men."

the aides-de-camp,

謁者, literally "visitors," is equivalent, as Tu Yu says, to 主告事者 "those whose duty it is to keep the general supplied with information," which naturally necessitates frequent interviews with him. Chang Yü goes too far afield for an explanation in saying that they are 典賓客之將 "the leaders of mercenary troops.".

the door-keepers and sentries

閽吏 and 守舍之人.

of the general in command.

守將, according to Chang Yü, is simply 守官任職之將 "a general on active service." Capt. Calthrop is wrong, I think, in making 守將 directly dependent on 姓名 (. . . "the names of the general in charge," etc.).

Our spies must be commissioned to ascertain these.

As the first step, no doubt, towards finding out if any of these important functionaries can be won over by bribery. Capt. Calthrop blunders badly with: "Then set the spies to watch them."

21. 必索敵人之間來間我者因而利之導而舍
之故反間可得而用也
22. 因是而知之故鄉間內間可得而使也
23. 因是而知之故死間爲誑事可使告敵
24. 因是而知之故生間有使如期

21. The enemy's spies who have come to spy on us must be sought out,

必索 is omitted by the *T'ung Tien* and *Yü Lan*. Its recurrence is certainly suspicious, though the sense may seem to gain by it. The *T'u Shu* has this variation: ... 敵間之來間吾者, etc.

tempted with bribes, led away and comfortably housed.

舍 is probably more than merely 居止 or 稽留 "detain." Cf. § 25 *ad fin.*, where Sun Tzŭ insists that these converted spies shall be treated well. Chang Yü's paraphrase is 館舍.

Thus they will become converted spies and available for our service.

22. It is through the information brought by the converted spy that we are able to acquire and employ local and inward spies.

Tu Yu expands 因是而知之 into 因反敵間而知敵情 "through conversion of the enemy's spies we learn the enemy's condition." And Chang Yü says: 因是反間知彼鄉人之貪利者官人之有隙者誘而使之 "We must tempt the converted spy into our service, because it is he that knows which of the local inhabitants are greedy of gain, and which of the officials are open to corruption." In the *T'ung Tien*, 鄉 has been altered to 因, doubtless for the sake of uniformity with § 9.

23. It is owing to his information, again, that we can cause the doomed spy to carry false tidings to the enemy.

"Because the converted spy knows how the enemy can best be deceived" (Chang Yü). The *T'ung Tien* text, followed by the *Yü Lan*, has here the obviously interpolated sentence 因是可得而攻也.

24. Lastly, it is by his information that the surviving spy can be used on appointed occasions.

Capt. Calthrop omits this sentence.

25. 五間之事主必知之知之必在於反間故反
間不可不厚也

26. 昔殷之興也伊摯在夏周之興也呂牙在殷

25. The end and aim of spying in all its five varieties is knowledge of the enemy;

I have ventured to differ in this place from those commentators — Tu Yu and Chang Yü — who understand 主 as 人主, and make 五間之事 the antecedent of 之 (the others ignoring the point altogether). It is plausible enough that Sun Tzŭ should require the ruler to be familiar with the methods of spying (though one would rather expect 將 "general" in place of 主). But this involves taking 知之 here in quite a different way from the 知之 immediately following, as also from those in the previous sentences. 之 there refers vaguely to the enemy or the enemy's condition, and in order to retain the same meaning here, I make 主 a verb, governed by 五間之事. Cf. XI. § 19, where 主 is used in exactly the same manner. The sole objection that I can see in the way of this interpretation is the fact that the 死間, or fourth variety of spy, does not add to our knowledge of the enemy, but only misinforms the enemy about us. This would be, however, but a trivial oversight on Sun Tzŭ's part, inasmuch as the "doomed spy" is in the strictest sense not to be reckoned as a spy at all. Capt. Calthrop, it is hardly necessary to remark, slurs over the whole difficulty.

and this knowledge can only be derived, in the first instance, from the converted spy.

As explained in §§ 22—24. He not only brings information himself, but makes it possible to use the other kinds of spy to advantage.

Hence it is essential that the converted spy be treated with the utmost liberality.

26. Of old, the rise of the Yin dynasty

Sun Tzŭ means the 商 Shang dynasty, founded in 1766 B.C. Its name was changed to Yin by 盤庚 P'an Kêng in 1401.

was due to I Chih

Better known as 伊尹 I Yin, the famous general and statesman who took part in Ch'êng T'ang's campaign against 桀癸 Chieh Kuei.

27. 故惟明君賢將能以上智爲間者必成大功
此兵之要三軍之所恃而動也

who had served under the Hsia. Likewise, the rise of
the Chou dynasty was due to Lü Ya

呂尙 Lü Shang, whose "style" was 子牙, rose to high office
under the tyrant 紂辛 Chou Hsin, whom he afterwards helped to overthrow. Popularly known as 太公, a title bestowed on him by Wên
Wang, he is said to have composed a treatise on war, erroneously identified with the 六韜.

who had served under the Yin.

There is less precision in the Chinese than I have thought it well to
introduce into my translation, and the commentaries on the passage are
by no means explicit. But, having regard to the context, we can hardly
doubt that Sun Tzŭ is holding up I Chih and Lü Ya as illustrious examples
of the converted spy, or something closely analogous. His suggestion is,
that the Hsia and Yin dynasties were upset owing to the intimate knowledge of their weaknesses and shortcomings which these former ministers
were able to impart to the other side. Mei Yao-ch'ên appears to resent
any such aspersion on these historic names: "I Yin and Lü Ya," he
says, "were not rebels against the Government (非叛於國也).
Hsia could not employ the former, hence Yin employed him. Yin could
not employ the latter, hence Chou employed him. Their great achievements were all for the good of the people." Ho Shih is also indignant:
伊呂聖人之耦豈爲人間哉今孫子引之者言
五間之用須上智之人如伊呂之才智者可以
用間蓋重之之辭耳 "How should two divinely inspired men
such as I and Lü have acted as common spies? Sun Tzŭ's mention of
them simply means that the proper use of the five classes of spies is a
matter which requires men of the highest mental calibre like I and Lü,
whose wisdom and capacity qualified them for the task. The above
words only emphasise this point." Ho Shih believes then that the two
heroes are mentioned on account of their supposed skill in the use of
spies. But this is very weak, as it leaves totally unexplained the significant
words 在夏 and 在殷. Capt. Calthrop speaks, rather strangely, of
"the province of Yin . . . the country of Hsia . . . the State of Chu . . .
the people of Shang."

27. Hence it is only the enlightened ruler and the wise
general who will use the highest intelligence of the army
for purposes of spying,

Ch'ên Hao compares § 15: 非聖智不能用間. He points out that 湯武之聖伊呂宜用 "the god-like wisdom of Ch'êng T'ang and Wu Wang led them to employ I Yin and Lü Shang." The *T'u Shu* omits 惟.

and thereby they achieve great results.

Tu Mu closes with a note of warning: 夫水所以能濟舟亦有因水而覆沒者間所以能成功亦有憑間而傾敗者 "Just as water, which carries a boat from bank to bank, may also be the means of sinking it, so reliance on spies, while productive of great results, is oft-times the cause of utter destruction."

Spies are a most important element in war, because on them depends an army's ability to move.

The antecedent to 此 must be either 間者 or 用間者 understood from the whole sentence. Chia Lin says that an army without spies is like a man without ears or eyes.

CHINESE CONCORDANCE

Ai	愛	VIII. 12; X. 25, 26; XI. 18, 66; XIII. 2.
„	阨	X. 21.
an	安	II. 20; V. 22; VI. 4; XII. 22.
Cha	詐	VII. 15.
ch'a	察	I. 2; VIII. 14; IX. 39; X. 13, 20; XI. 41.
chan	戰	passim.
„	霑	XI. 28.
chang	障	IX. 21.
„	仗	IX. 29.
ch'ang	常	VI. 32, 34; X. 18; XI. 29.*
„	長	VI. 34.
„	嘗	V. 9.
chao	朝	VII. 28.
chê	者	passim.
„	折	V. 13; XI. 63.
chên	軫	XII. 4.*
ch'ên	陳	VII. 32; IX. 25, 27; X. 18.
„	塵	IX. 23.
chêng	爭	III. 7; VII. 3, 5, 6, 7, 9, 10, 22; VIII. 3; XI. 1, 4, 11, 47, 55; XIII. 2.
chêng	正	V. 3, 5, 10, 11; VII. 32; XI. 35.
„	政	III. 3, 14; IV. 16; VII. 23; XI. 32, 56, 63.
„	整	XI. 18.
ch'êng	成	III. 4; XI. 62; XIII. 4, 27.
„	城	II. 2; III. 3, 4, 5, 6; VIII. 3; XI. 7, 55; XIII. 20.
„	乘²	II. 4, 17; XI. 19.
„	乘⁴	II. 1, 17.
„	稱	IV. 17, 18, 19.
chi	計	I. 3, 12, 15, 16; VI. 22; VII. 4, 22; X. 21; XI. 22.
„	及	VI. 10; VII. 6; XI. 15, 19, 68.
„	汲	IX. 30.
„	急	II. 12.
„	己	III. 18; IV. 2; VI. 18; X. 31; XI. 55.
„	紀	XIII. 8.
„	惎	II. 15.
„	擊	VI. 15, 30; VII. 29, 32; VIII. 3; IX. 4; X. 7, 15, 19, 27, 28, 29; XI. 9, 29; XIII. 20.

chi 巫 IX. 7, 15; XI. 65.

„ 極 VI. 25; XII. 8.

„ 集 IX. 32; XI. 16.

„ 激 V. 12.

„ 既 III. 16; VII. 25.

„ 疾 V, 12, 13; VII. 17; IX. 12; XI. 10.

„ 機 V. 15; XI. 38.

„ 飢 VI. 4; VII. 31; IX. 29.

„ 積 IV. 20; VII. 11; XI. 22; XII. 1.

„ 戟 II. 14.

„ 籍 II. 8.

„ 箕 XII. 4.*

„ 濟 IX. 4; XI. 30.

„ 繼 XI. 49.

ch'i 其 passim.

„ 期 IX. 27; XI. 38, 66; XIII. 24.

„ 旗 II. 17; VII. 23, 24, 26, 32; IX. 33.

„ 器 III. 4.

„ 漆 II. 1.

„ 起 II. 4; IX. 22; XII. 3, 4; XIII. 8.

„ 隙 III. 11; IX. 15.

„ 齊 IX. 43; XI. 16, 32.

„ 七 II. 13; XIII. 1.

„ 奇 V. 3, 5, 6, 10, 11.

„ 谿 IV. 20; X. 25.

„ 氣 VII. 27, 28, 29; XI. 22.

chia 家 I. 25; II. 13, 14, 20; XIII. 1.

„ 甲 II. 1, 14; VII. 7.

„ 加 V. 4; XI. 54, 55.

chia 葭 IX. 17.

chiang 江 V. 6.

„ 疆 V. 17, 18.

„ 將¹ XI. 18, 46, 47, 48, 49, 50.

„ 將⁴ I. 4, 9, 11, 13, 15; II. 15, 20; III. 5, 11, 17; VII. 1, 7, 9, 27; VIII. 1, 4, 5, 12, 13, 14; IX. 33; X. 13, 14, 17, 18, 19, 20, 21; XI. 35, 40, 61; XII. 16, 18, 22; XIII. 3, 4, 20, 27.

„ 蔣 IX. 17.

ch'iang 强 I. 13, 21; II. 18; III. 11; IX. 24; X. 16, 19; XII. 13.

chiao 交 III. 3; VII. 2, 12; VIII. 2; IX. 8; XI. 1, 5, 12, 28, 48, 52, 54, 55.

„ 校 I. 3, 12.

„ 教 IX. 44; X. 18.

„ 驕 I. 22; X. 26.

„ 膠 II. 1.

ch'iao 巧 II. 5; XI. 62.

„ 樵 IX. 23.

chieh 竭 II. 11, 12; V. 6.

„ 皆 VI. 27; XI. 33; XIII. 19.

„ 戒 XI. 19, 25.

„ 潔 VIII. 12.

„ 節 V. 13, 14, 15.

„ 解 VIII. 9.

„ 結 XI. 48.

ch'ieh 且 III. 16; XI. 23.

ch'ieh 怯 V. 17, 18; VII. 25.

chien 閒 or 間 VI. 25; XIII. *passim.*

" 澗 IX. 15.

" 兼 VII. 7.

" 姦 IX. 17.

" 堅 III. 10.

" 賤 IX. 11; XI. 15.

" 踐 XI. 67.

" 見 I. 26; IV. 8, 10; VII. 23; IX. 31.

ch'ien 千 II. 1; IV. 20; V. 23; VI. 6, 19; XI. 61; XIII. 1.

" 淺 XI. 42, 44.

" 前 VI. 17, 20; IX. 9; XI. 15, 45.

chih 知 *passim.*

" 智 I. 9; II. 4, 15; IV. 12; VIII. 7; XIII. 15, 27.

" 之 *passim.*

" 之 [= 至] VI. 12; XI. 39.

" 止 V. 22; XI. 11, 17; XII. 8, 11, 19.

" 支 X. 1, 6, 7.

" 直 VII. 3, 4, 22.

" 制 I. 7, 10, 17; VI. 27, 31; X. 21.

" 志 XI. 46.

" 摯 XIII. 26.*

" 鷙 V. 13.

" 治 V. 1, 17, 18; VII. 29, 30, 31, 32; VIII. 6; X. 26; XI. 35.

" 至 III. 16; V. 12, 13; VI. 3, 9, 25; VII. 4, 8, 9, 10; IX. 14, 37; X. 13, 20; XI. 6, 26, 29; XIII. 2.

chih 致 VI. 2; XII. 18.

ch'ih 馳 II. 1.

" 斥 IX. 7, 8.

chin 近 I. 8, 19; II. 11; VI. 20; VII. 31; IX. 15, 16, 18; X. 21.

" 進 III. 13; VI. 10; VII. 25; IX. 19, 24, 28, 31, 40; X. 24; XI. 49.

" 盡 II. 7; XI. 23.

" 金 II. 1; VII. 23, 24; XIII. 1, 2.

" 謹 IX. 17, 39; XI. 22, 48.

" 禁 XI. 26.

" 襟 XI. 28.

ch'in 親 I. 23; IX. 42; XI. 25; XIII. 14.

" 擒 III. 10; VII. 7; IX. 41.

" 侵 VII. 18.

ching 靜 V. 22; VI. 23; VII. 30; IX. 18; XI. 35; XII. 7.

" 旌 II. 17: VII. 23, 24, 26; IX. 33.

" 井 IX. 15, 17.

" 勁 VII. 8.

" 經 I. 3.

" 精 IX. 37.

" 警 XII. 22.

" 境 XI. 43.

ch'ing 情 I. 3, 12; XI. 19, 41, 51; XIII. 2, 6.

" 請 IX. 26.

ch'ing 輕 IX. 25; XI. 1, 3, 11, 44, 46.

chio 角 VI. 24.

„ 爵 XIII. 2.

chiu 九 IV. 7; VIII. 4, 5, 6; XI. 41.

„ 久 II. 2, 3, 5, 6, 19; III. 6; IX. 39; XII. 11.

„ 救 VI. 11, 20; XI. 15, 30.

ch'iu 求 IV. 15; V. 21; X. 24; XI. 25.

„ 丘 II. 12, 14.

„ 邱 VII. 33; IX. 13.

„ 秋 IV. 10.

chiung 煢 IX. 36.

ch'iung 窮 V. 6, 10, 11; VI. 28; VII. 36; IX. 34; X. 30.

cho 拙 II. 5.

chou 畫 VII. 26, 28; XII. 11.

„ 舟 XI. 30, 39.

„ 周 III. 11; XIII. 26.*

„ 胄 II. 14.

chu 主 I. 10, 13; II. 20; X. 23, 24; XI. 19, 20; XII. 16, 18; XIII. 3, 25.

„ 諸 II. 4; III. 16; VII. 12; VIII. 10; XI. 2, 6, 28,* 38, 52.

„ 著 IX. 45.

„ 助 IX. 13; X. 21.

„ 誅 XI. 64.

„ 屬 XI. 6, 46.

ch'u 處 ³ VI. 1, 24, 30; VII. 7; IX. 1, 2, 6, 8, 9, 12, 13; XI. 68.

ch'u 處 ⁴ IX. 17.

„ 出 I. 24; V. 6; VI. 5; IX. 25; X. 5, 6, 7; XIII. 1, 4.

chuan 專 VI. 13, 14; VII. 25; XI. 20, 42.

„ 轉 V. 22, 23.

ch'uan 傳 I. 25; XIII. 12.

chui 追 VI. 10.

„ 隊 XII. 1.

chun 諄 IX. 35.

chung 眾 passim.

„ 重 VII. 6, 11; IX. 33; XI. 1, 7, 13, 44, 49.

„ 鍾 II. 15.

„ 終 V. 6.

„ 中 II. 13; IX. 8; XI. 29.

ch'ung 衝 VI. 10.

chü 居 IX. 20, 25; X. 3, 8, 9, 10, 11; XI. 37.

„ 舉 II. 1; IV. 10; VII. 6; X. 30; XI. 63.

„ 聚 VII. 2; VIII. 1; XI. 40, 54.

„ 車 II. 1, 14, 17; IX. 23, 25.

„ 具 III. 4; XII. 2.

„ 俱 X. 25; XI. 29; XIII. 8.

„ 沮 VII. 13; XI. 8, 52.

„ 拒 XI. 68.

„ 距 III. 4.

„ 拘 XI. 24.

„ 懼 XI. 24.

ch'ü 去 I. 15; II. 13, 14; IX. 7, 15, 39; X. 7, 11; XI. 26, 38, 43.

ch'ü　取　I. 20; II. 9, 16; V. 19;
　　　　VI. 7, 33; IX. 40, 43;
　　　　XII. 15; XIII. 5, 6.

,,　　屈　II. 2, 4, 13; III. 2, 6;
　　　　VIII. 10; XI. 41.

,,　　趨　VI. 1, 5, 29, 30; VII.
　　　　7; VIII. 10; XI. 47.

,,　　驅　IX. 24; XI. 39.

,,　　衢　VIII. 2; XI. 1, 6, 12,
　　　　43, 48.

,,　　曲　I. 10.

chüan　卷　VII. 7.

,,　　倦　IX. 33.

ch'üan　全　III. 1, 7; IV. 7; X. 31;
　　　　XII. 22.

,,　　權　I. 17; III. 15; VII. 21;
　　　　XI. 55.

chüeh　絶　VIII. 2; IX, 1, 3, 4, 7,
　　　　15; XI. 22, 43; XII. 14.

,,　　決　IV. 20; XI. 67.

,,　　蹶　VII. 9.

ch'üeh　闕　VII. 36; XI. 50.

chün　君　III. 12, 17; VII 1; VIII.
　　　　1, 3; XII. 22; XIII.
　　　　4, 8, 27.

,,　　軍　passim.

,,　　均　X. 12, 15.

ch'ün　羣　XI. 39.

Êrh　二　I. 4; II. 15; IV. 17;
　　　　VII. 10; XII. 1.

êrh　耳　IV. 10; VII. 24, 26;
　　　　XI. 36.

,,　　兒　X. 25.

,,　　而　passim.

Fa　法　I. 4, 10, 13; II. 1; III.
　　　　1, 4, 7, 8; IV. 16, 17;

VII. 1, 8, 9, 22, 25,
33, 37; VIII. 1, 11;
XI. 1, 56.

Fa　發　V. 15; VII. 4; XI. 28, 38;
　　　　XII. 3, 6, 7, 9, 10;
　　　　XIII. 19.

,,　　罰　I. 13; IX. 36, 42.

,,　　伐　III. 3; XI. 54.

fan　反　XIII. 7, 11, 13, 21, 25.

,,　　返　IX. 34; X. 4, 5.

,,　　凡　passim.

,,　　犯　XI. 56, 57.

,,　　煩　VIII. 12.

fang　方　V. 22; XI. 31.

,,　　防　IX. 13.

fei　費　II. 1, 13, 14; XII. 15;
　　　　XIII. 1.

,,　　非　III. 2, 6; IV. 8, 9; IX. 40;
　　　　X. 14; XI. 27, 53; XII.
　　　　17; XIII. 3, 15, 16, 17.

fên　分　III. 5, 8; V. 1; VI. 13, 14;
　　　　VII. 10, 16, 20.

,,　　忿　III. 5; VIII. 12.

,,　　紛　V. 16.

,,　　焚　XI. 39.

,,　　轒　III. 4.

fêng　風　VII. 17; XI. 30; XII. 4,
　　　　10, 11.

,,　　奉　II. 1; XIII. 1.

,,　　鋒　X. 19.

fou　瓿　IX. 34.

,,　　覆　IX. 22.

fu　符　XI. 63.

,,　　附　III. 5; IX. 5, 42.

fu 夫 *passim.*

„ 扶 XI. 15.

„ 復 V. 6; VI. 28; XII. 20, 21.

„ 覆 VIII. 14; IX. 17.

„ 伏 IX. 17, 22.

„ 負 I. 14, 26; III. 18.

„ 服 IX. 42, 44; X. 17.

„ 釜 XI. 39.

„ 赴 X. 25.

„ 輔 III. 11.

Hai 害 II. 7; VI. 3; VIII. 7, 9, 10; XI. 57, 59.

han 寒 I. 7.

hao 毫 IV. 10.

hêng 橫 X. 18.

ho 合 V. 5; VII. 2, 16; VIII. 1, 2; IX. 39; X. 19, 24; XI. 12, 16, 17, 54; XII. 19.

„ 闔 XI. 65.

„ 何 XI. 18.

„ 河 V. 6.

„ 和 VII. 2; IX. 26.

hou 厚 X. 26; XIII 14, 25.

„ 侯 II. 4; III. 16; VII. 12; VIII. 10; XI. 2, 6, 52.

„ 後 *passim.*

hsi 昔 IV. 1; XIII. 26.

„ 喜 IX. 11; XII. 20.

„ 奚 VI. 21.

„ 翕 IX. 35.

„ 息 IX. 38.

„ 攜 XI. 34.

hsia 下 III. 3, 7, 17; IV. 7, 9; VI. 29; IX. 11; XI. 6, 15, 55; XII. 10.

„ 夏 XIII. 26.*

hsia 狹 I. 8.

hsiang 相¹ V. 11; VII. 23; IX. 39, 45; XI. 15, 30; XIII. 2.

„ 相⁴ IX. 1.

„ 鄉 VII. 14, 20; XI. 52; XIII. 7, 9, 22.

„ 向 VII. 33; XI. 61.

„ 象 VI. 29; XIII. 5.

„ 祥 XI. 26.

„ 詳 XI. 60.

hsiao 小 III. 10; IX. 17.

hsieh 械 III. 4.

„ 駭 IX. 22.

„ 謝 IX. 38.

hsien 先 *passim.*

„ 險 I. 8; V. 14; VII. 13; IX. 17, 18; X. 1, 10, 21; XI. 8, 40, 52.

„ 陷 IX. 15; X. 14, 16; XI. 24, 58, 59.

hsien 賢 XIII. 4, 27.

hsin 信 I. 9; IX. 45; XI. 25.

„ 心 VII. 27, 30.

hsing 行 I. 13; V. 22; VI. 6, 29, 34; VII. 7, 13; IX. 42, 44; XI. 8, 13, 52; XII. 2.

„ 形 *passim.*

„ 興 XII. 18; XIII. 1, 26.

„ 性 V. 22.

hsing	姓	II. 10, 11, 13; XIII. 1, 20.
hsiu	修	III. 4; IV. 6; XI. 25; XII. 15, 16.
"	休	IX. 38.
hsiung	凶	XII. 15.
hsü	虛	II. 13; V. 4; VI. 10; IX. 32.
"	徐	VII. 17; IX. 35.
"	宿	XII. 4.
hsüan	懸	VII. 21; IX. 34; XI. 56.
"	選	X. 19.
hsün	循	V. 11.
hu	乎	I. 26; VI. 9; XI. 30.
"	呼	IX. 32.
"	戶	XI. 68.
hua	化	VI. 33.
"	畫	VI. 12.
"	譁	VII. 30.
huan	患	III. 12; VII. 3; VIII. 9.
"	環	V. 11.
huang	黃	IX. 10.*
"	潢	IX. 17.
hui	毀	III. 6; V. 13.
"	隳	XI. 55.
"	會	VI. 19.
hun	渾	V. 16.
huo	貨	II. 4, 16; XI. 27.
"	火	VII. 18, 26; XII. passim.
"	惑	III. 14, 16.
"	活	XI. 50.
I	一	passim.
"	已	II. 17; III. 4; IV. 13; IX. 40, 42; XI. 24, 34, 51.
I	易	I. 8; IV. 11; IX. 9, 20, 41; XI. 37.
"	意	I. 5, 24; VI. 5; XI. 60.
"	益	II. 18; VI. 21; IX. 24, 40.
"	鎰	IV. 19.
"	疑	III. 15, 16; IX. 21; XI. 26.
"	佚	I. 23; VI. 1, 4; VII. 31.
"	役	II. 8, 12; VIII. 10.
"	亦	VI. 21; XI. 4.
"	依	IX. 1, 8.
"	倚	IX. 29.
"	伊	XIII. 26.*
"	邑	XI. 7.
i	頤	XI. 28.
"	夷	XI. 63.
"	義	XIII. 16.
"	蟻	III. 5.
"	翼	XII. 4.*
"	鷁	IX. 17.
"	以	passim.
"	矣	passim.
Jan	然	II. 1; XI. 29, 30, 58, 59.
jao	擾	IX. 33.
"	饒	XI. 21.
jên	人	passim.
"	仁	I. 9; XIII. 2, 16.
"	任	III. 15; V. 21, 22; X. 13, 20.
"	仞	IV. 20; V. 23.
jih	日	II. 1; IV. 10; V. 6; VI. 19, 20, 34; VII. 7; XI. 28, 63; XII. 3, 4; XIII. 1, 2.

jo 若 III. 9; IV. 19, 20; IX. 8; X. 5, 9, 11; XI. 18, 32, 34, 39, 56.

„ 弱 III. 11; V. 17, 18; X. 16, 18, 19.

jou 肉 IX. 34.

„ 柔 XI. 33.

ju 辱 VIII. 12.

„ 入 IX. 35; XI. passim.

„ 如 V passim; VII. 17, 18, 19; X. 25, 26; XI. 29, 30, 38, 68; XIII. 24.

jui 銳 II. 2, 4; VII. 28, 29, 34; IX. 23.

K'ai 開 XI. 65, 68.

kan 敢 XI. 18, 30.

„ 秆 II. 15.

kang 剛 XI. 33.

kao 高 VI. 11, 29; VII. 33; IX. 2, 6, 9, 11, 23; X. 3, 10; XI. 38.

„ 告 XI. 57; XIII. 19, 23.

kêng 更 II. 17.

ko 革 II. 1; XI. 37.

k'o 渴 IX. 30.

„ 客 II. 1; IX. 4, 5; XI. 20, 42.

„ 克 XI. 20.

„ 可 passim.

kou 溝 VI. 11.

k'ou 寇 VII. 36; IX. 34.

ku 古 IV. 11; XI. 15.

„ 固 VI. 7; XI. 24. 45, 48.

„ 故 passim.

„ 谷 IX. 1.

ku 鼓 VII. 23, 24, 26.

k'u 庫 XII. 1.

kua 寡 III. 17; V. 1, 2; VI. 14, 15, 16, 17, 18; XI. 9, 15.

„ 挂 X. 1, 4, 5.

kuai 乖 VI. 12.

kuan 官 I. 10; XIII. 10.

„ 關 XI. 63.

„ 觀 I. 26; V. 8.

kuang 廣 I. 8; IX. 23.

k'uang 況 I. 26; VI. 20.

„ 誑 XII. 12, 23.

kuei 歸 VII. 28, 29, 35; XI. 9.

„ 鬼 XIII. 5.

„ 貴 II. 11, 19; IX. 11; XI. 15.

„ 劌 XI. 28.*

„ 詭 I. 18.

k'uei 窺 VI. 25.

„ 饋 II. 1.

k'un 困 IX. 36.

kung 公 II. 14; XIII. 1.

kung 功 IV. 12; XIII. 4, 27.

„ 攻 passim.

„ 共 VI. 14.

k'ung 恐 IX. 32.

kuo 國 I. 1; II. 3, 6, 9, 10, 20; III. 1, 6, 11; X. 24; XI. 43, 54, 55; XII. 21, 22.

„ 過 IV. 8; V. 7, 8, 9, 10; VIII. 13; X. 14; XI. 51.

„ 彊 V. 15.

k'uo 廓 VII. 20.

Lai 來 VIII. 11; IX. 4, 21, 23, 38; X. 2; XI. 5, 18, 39; XIII. 21.

lang 廊 XI. 64.

lao 勞 I. 23; VI. 1, 4, 6; VII. 31; IX. 31; XI. 22.

„ 牢 IX. 15.

lei 壘 VI. 11.

„ 雷 IV. 10; VII. 19.

li 吏 IX. 33; X. 16, 17, 18.

„ 里 II. 1; VI. 6, 19, 20; VII. 7, 9, 10; XI. 61; XIII. 1.

„ 理 VI. 23; XI. 33, 41.

„ 力 II. 2, 4, 13; IV. 10; VII. 31; IX. 40; XI. 22, 23; XII. 8.

„ 立 IV. 14; VII. 15; IX. 29.

„ 離 I. 23; XI. 16.

„ 厲 XI. 64.

„ 利 passim.

liang 量 IV. 17, 18.

„ 糧 II. 1, 8, 9; VII. 11; X. 3.

„ 良 XII. 16, 22.

liao 料 IX. 40; X. 19, 21.

lien 廉 VIII. 12.

„ 練 I. 13.

lin 林 VII. 13, 17; IX. 17; XI. 8, 52.

ling 令 I. 5, 13; IX. 4, 43, 44, 45; X. 7, 26; XI. 25, 28, 56; XIII. 12, 20.

„ 陵 VII. 33; IX. 13.

liu 六 II. 14; X. 13, 14, 20.

liu 留 I. 15; VIII. 2; IX. 7; XII. 15.

„ 流 VI. 31; IX. 6.

lo 羅 IX. 15.

lu 虜 VIII. 12.

„ 櫓 II. 14; III. 4.

„ 路 XIII. 1.

„ 陸 IX. 9.

„ 祿 XIII. 2.

luan 亂 I. 20; III. 16; V. 16, 17, 18; VII. 30; IX. 33; X. 14, 18, 26.

„ 卵 V. 4.

lun 輪 XI. 31.

lung 隆 IX. 2.

lü 慮 VIII. 7; IX. 41; XI. 37; XII. 16.

„ 呂 XIII. 26.*

„ 旅 III. 1.

„ 屢 IX. 36.

lüeh 掠 VII. 18, 20; XI. 13, 21.

Ma 馬 II. 14; IX. 34; XI. 31.

mai 賣 II. 11.

„ 埋 XI. 31.

mei 每 III. 18.

mên 門 XIII. 20.

mi 麋 III. 13.

„ 迷 X. 30.

„ 密 XIII. 14.

miao 廟 I. 26; XI. 64.

„ 妙 XIII. 17.

min 民 I. 5, 6; II. 20; IV. 20; VII. 24, 25, 26; VIII. 12; IX. 44; X. 24.

ming 命 II. 20; VI. 9; VII. 1; VIII.
  1, 3; XI 27; XII. 15.

" 名 IV. 12; V. 2; X. 24;
  XIII. 20.

ming 明 I. 13; IV. 10; X. 18; XII.
  13, 16, 22; XIII. 4, 27.

mo 沫 IX. 14.

" 墨 XI. 67.

" 莫 I. 11; VI. 27; VII. 3; XI.
  39; XIII. 8, 14.

mou 謀 III. 3, 7; VI. 25; VII. 12;
  VIII. 2; IX. 26; XI. 14,
  22, 37, 52.

mu 目 IV. 10; VII. 24, 26; XI. 36.

" 木 V. 22.

" 暮 VII. 28.

Nai 乃 I. 16; X. 31.

nan 難 III. 16; VII. 3, 19; IX.
  42; X. 4, 5, 12; XI. 8.

nao 橈 I. 22.

nei 內 II. 1, 13; IX. 4; XII. 6,
  9; XIII. 1, 7, 10, 22.

nêng 能 passim.

ni 餌 VII. 35.

" 逆 VII. 33.

niao 鳥 V. 13; IX. 22, 32.

nien 年 XIII. 2.

niu 牛 II. 14.

nu 怒 I. 22; II. 16; IX. 33, 39;
  X. 17; XII. 18, 20.

" 弩 II. 14; V. 15.

nü 女 XI. 68.

O 遏 VII. 35.

Pa 拔 III. 5, 6; XI. 55.

pa 覇 XI. 53, 54.

pai 敗 I. 15; IV. 13, 14, 15, 16,
  19; V. 3, 16; VI. 21;
  X. 20, 22; XI. 59.

pan 半 VII. 9; IX. 4, 28; X. 7,
  27, 28, 29.

p'ang 旁 IX. 17.

pao 保 IV. 7, 16; X. 24.

" 寶 X. 24; XIII. 8.

" 報 XIII. 13.

" 暴 II. 3; IX. 37.

" 飽 VI. 4; VII. 31.

pei 倍 III. 8; VII. 7.

" 北 VII. 34; X. 14, 19; XI. 23.

" 背 VII. 33; IX. 8, 9, 13, 16;
  XI. 7, 45.

" 卑 I. 22; IX. 23, 24.

" 備 I. 21, 24; VI. 16, 17, 18;
  IX. 24; X. 5.

" 奔 pên IX. 27.

p'êng 崩 X. 14, 17.

pi 壁 XII. 4.*

" 避 I. 21; III. 9; VI. 29; VII.
  29; X. 24.

" 弊 II. 4.

" 蔽 II. 14.

" 必 passim.

" 彼 III. 18; X. 2, 6, 31; XI.
  4, 5, 9.

p'i 譬 X. 26; XI. 29.

" 圮 VIII. 2; XI. 1, 8, 13, 49.

" 罷 II. 14; VII. 8.

p'iao 漂 V. 12.

13

pien 變 V. 7, 8, 9, 10; VI. 33;
VII. 16, 26, 32; VIII.
4, 5, 6; XI. 41; XII. 5, 12.

pin 賓 II. 1.

p'in 貧 II. 10.

ping 并 XI. 61.

„ 併 IX. 40; XI. 22.

„ 兵 *passim.*

p'ing 平 IX. 9.

po 百 II. 10, 11, 13; III. 2, 18;
VII. 7; IX. 12; XIII.
1, 2.

„ 迫 VII. 36.

p'o 破 II. 14; III. 1; XI. 39.

pu 不 *passim.*

Sai 塞 XI. 50.

san 三 I. 4; II. 8; III. *passim;*
IV. 17; V. 2; VII. 7,
10, 27; XI. 6, 21, 40,
56; XII. 1; XIII. 14, 27.

„ 散 IX. 23; XI. 1, 2, 11, 42, 46.

sao 燥 XII. 4.

„ 驕 XIII. 1.

sê 色 V. 8.

sha 殺 II. 16; III. 5; VIII. 12,
14; XI. 61; XIII. 20.

shan 山 V. 23; VII. 13, 18; IX.
1, 2; XI. 8, 29, 52.

„ 善 *passim.*

shang 上 I. 5; III. 1, 3, 17; IV. 7;
VII. 9; IX. 6, 14; X.
21; XI. 15, 64; XII. 10;
XIII. 27.

„ 賞 I. 13; II. 17; IX. 36; XI.
56; XIII. 14.

shao 少 I. 26; III. 9; IX. 23; X. 19.

shê 舍 VII. 2; VIII. 2; IX. 34;
XIII. 20, 21.

„ 蛇 XI. 29.

„ 涉 IX. 14.

shên 深 VI. 11, 25; X. 25; XI.
passim.

„ 信 VIII. 8; XI. 55. [See under
hsin.]

„ 伸 XI. 41.

„ 神 VI. 9, 33; XIII. 5, 8.

„ 甚 XI. 24.

„ 慎 XII. 22.

shêng 勝 *passim.*

„ 生 I. 2, 6, 8; IV. 18; V. 6,
11, 17; VI. 23, 34; VIII.
12; IX. 2, 6, 9, 12, 17;
XI. 58; XII. 21; XIII.
7, 13, 24.

„ 聲 V. 7; VI. 9.

„ 聖 XIII. 15.

shih 是 *passim.*

„ 矢 II. 14.

„ 失 IV. 14; VI. 22; IX. 35.

„ 石 II. 15; V. 12, 22, 23.

„ 始 V. 6; XI. 68.

„ 示 I. 19; XI. 50.

„ 施 XI. 56.

„ 弛 X. 14, 16.

„ 時 I. 7; V. 6; VI. 34; XII.
3, 4, 9.

„ 識 III. 17; XI. 37.

„ 執 V. *passim.*

„ 勢 I. 16, 17; VI. 32; X. 12, 15.

shih 十 II. 1, 13, 14, 15, 17; III. 8; VI. 14, 20; VII. 8, 9, 10; X. 15; XIII. 1.

„ 士 I. 13; III. 5, 14, 15; XI. 23, 24, 27, 28, 36.

„ 寶 I. 21; V. 4; VI. 30; IX. 12; XIII. 17.

„ 使 IV. 3; V. 3; VI. 3, 18, 22; X. 26; XI. passim; XIII. 16, 22, 23, 24.

„ 事 I. 1; III. 14; XI, XIII, passim.

„ 恃 VIII. 11; IX. 18; XI. 15, 31; XIII. 27.

„ 師 II. 1, 3, 10, 11; VII. 36; XI. 43; XII. 18; XIII. 1.

„ 視 VII. 23; IX. 2, 6; X. 25.

„ 食 II. 9, 15; VII. 11, 35; IX. 34; XI. 21, 49.

shou 受 V. 3; VII. 1; VIII. 1, 3.

„ 守 IV. 5, 6, 7; VI. 7, 8, 12; XI. 48; XII. 12; XIII. 2, 20.

„ 手 XI. 30, 34.

„ 獸 IX. 22.

„ 首 XI. 29.

„ 壽 XI. 27.

shu 數 IV. 17, 18; V. 1, 18; VI. 20; IX. 36; XII. 12; XIII. 2.

„ 樹 IX. 8, 21.

„ 孰 I. 13; V. 11.

„ 銖 IV. 19.

„ 輸 II. 10.

„ 暑 I. 7.

„ 術 VIII. 6.

shuai 率 XI. 29, 30.

„ 帥 XI. 38.

shui 水 IV. 20; V. 12; VI. 29, 31, 32; IX. 3, 4, 5, 6, 8, 14; XII. 13, 14.

shun 楯 II. 14.

„ 順 XI. 60.

so 所 passim.

„ 索 I. 3, 12; IX. 17; XIII. 20, 21.

ssŭ 死 I. 2, 6, 8; V. 6; VI. 23, 34; VIII. 2, 12; IX. 9, 10; X. 25; XI. passim; XII. 21; XIII. 7, 12, 19, 23.

„ 四 I. 4; IV. 17; V. 6; VI. 34; XI. 43, 53; XII. 1, 4.

„ 駟 II. 1.

„ 司 II. 20; VI. 9.

„ 私 XI. 55.

su 速 II. 5; VI. 10; VIII. 12; XI. 19.

„ 素 IX. 44, 45; XII. 2.

„ 粟 IX. 34.

suan 算 I. 26.

sui 雖 II. 4; VI. 11, 21, 22; VIII. 5, 6; X. 7.

„ 隨 XI. 67.

sun 孫 passim.*

Ta 大 I. 1; II. 14; III. 10; X. 17; XI. 54; XIII. 27.

„ 達 IX. 23; XI. 43.

tai 待 III. 17; IV. 1; V. 20; VI. 1; VII. 30, 31; VIII. 11; IX. 14; X. 8, 10; XI. 18; XII 7, 9.

tai	殆	III. 18; X. 31.
„	怠	XIII. 1.
„	帶	II. 1.
tan	殫	II. 4, 13.
tang	當	II. 15; XI. 30.
t'ang	堂	VII. 32.
tao	道	I. *passim;* III. 17; IV. 16; VII. 7; X. *passim;* XI. 8, 19, 20, 32, 42; XII. 22; XIII. 1, 8.
„	導	VII. 14; XI 52; XIII. 21.
t'ao	逃	III. 9.
tê	得	*passim.*
t'ê	忒	IV. 13.
têng	登	IX. 2; XI. 38.
ti	地	I. 2, 4, 8, 13; IV. 7, 14, 18; V. 6; VI. *passim;* VII. 14, 20; VIII. 2, 3, 5; IX. 13, 15; X. 1, 13, 21, 29, 31; XI. *passim.*
„	敵	II. 9, 15, 16, 18; III. 9, 10; IV. 1, 2, 3, 14; V. 3, 19; VI, IX, X, XI, XIII. *passim.*
„	帝	IX. 10.
„	隄	IX. 13.
t'i	梯	XI. 38.
„	涕	XI. 28.
t'iao	桃	IX. 19; X. 12.
„	條	IX. 23.
t'ien	天	I. 4, 7, 13; III. 7; IV. 7, 9; V. 6; IX. 15; X. 14, 31; XI. 6, 55; XII. 4.
ting	定	IX. 14.

t'ing	聽	I. 15, 16; V. 7; XI. 18.
„	霆	IV. 10; VII. 19.
to	度	IV. 18; VI. 21; XIII. 5.
„	惰	VII. 28, 29.
„	奪	VII. 27; XI. 18; XII. 14.
„	多	I. 26; IV. 10; VI. 16, 21; VII. 26; IX. 21, 40; XI. 7.
t'o	脫	XI. 68.
tou	鬥	V. 2, 16; VI. 22; XI. 24, 51.
t'ou	投	V. 4; XI. 23, 28, 40, 58.
tsa	雜	II. 17; VIII. 7, 8, 9.
tsai	在	*passim.*
„	災	III. 5; VIII. 13; X. 14; XI. 26.
„	哉	VI. 21; XIII. 18.
„	再	II. 8.
„	載	II. 8.
ts'ai	財	II. 11, 12, 13; XI. 27.
„	材	II. 1.
„	探	IX. 23.
ts'ang	藏	IV. 7; IX. 17.
tsao	早	XII. 6.
ts'ao	草	IX. 8, 21.
„	操	XIII. 1.
tsê	則	*passim.*
„	擇	V. 21.
„	澤	VII. 13; IX. 7, 8; XI. 8, 52.
„	責	V. 21.
ts'ê	側	IX. 25.
„	測	XI. 22.

ts'ê 策 VI. 22.

tso 左 VI. 17, 20; XI. 30; XIII. 20.

„ 佐 I. 16; XII. 13; XIII. 3.

„ 作 VI. 23.

„ 坐 XI. 28.

ts'o 挫 II. 2, 4.

„ 措 IV. 13.

„ 錯 VI. 26.

tsou 走 IX. 27; X. 14, 15.

tsu 卒 I. 13; II. 17; III. 1; V. 20; VII. 34; IX. 42; X. 16, 18, 25, 27, 28, 29; XI. 16, 28, 36.

„ 足 II. 3, 9; IV. 6; VI. 24; IX. 40; XI. 21, 31.

„ 阻 VII. 13; IX. 17; XI. 8, 52.

tsui 罪 X. 24.

ts'un 存 I. 2; XI. 10, 58; XII. 21.

tsung 縱 X. 18.

ts'ung 從 V. 19; VII. 34; X. 9, 11; XI. 9, 51; XII. 8.

„ 聰 IV. 10.

tu 睹 II. 5.

„ 獨 VII. 25.

t'u 徒 IX. 23.

„ 兎 XI. 68.

„ 途 VII. 4; XI. 37.

„ 塗 VIII. 3; XI. 49.

tuan 短 V. 14; VI. 34.

„ 端 V. 11.

„ 碬 V. 4.

tui 懟 X. 17.

t'ui 退 III. 13; VI. 10; VII. 25; IX. 24, 28; X. 24.

tun 沌 V. 16.

„ 鈍 II. 2, 4.

„ 頓 III. 7.

tung 動 IV. 7; V. 19, 20, 22; VI. 4, 23; VII. 15, 18, 19, 21; IX. 21, 33; X. 30; XI. 17; XII. 17, 19; XIII. 1, 4, 27.

t'ung 通 VIII. 4, 5; X. 1, 2, 3; XI. 63.

„ 同 I. 5; III. 14, 15, 17; XI. 30.

tzŭ 子 I. 1; X. 25, 26; et al.

„ 自 IV. 7; VI. 3; X. 17; XI. 2.

„ 輜 VII. 6, 11; XII. 1.

tz'ŭ 此 passim.

Wai 外 I. 16; II. 1; XII. 6, 9; XIII. 1, 12.

wan 萬 II. 1; XIII. 1.

wang 往 IX. 23; X. 2, 4; XI. 5, 23, 24, 28, 39, 45.

„ 亡 I. 2; VII. 11; XI. 10, 58; XII. 21.

„ 王 XI. 53, 54.

wei 爲 passim.

„ 謂 II. 18; III. 13, 16; IV. 11; VI. 33; IX. 12, 43; XI. 15, 40, 62; XIII. 8.

„ 畏 I. 6; IX. 37.

„ 危 I. 6; II. 20; V. 22; VII. 5; VIII. 12, 14; XII. 17.

„ 唯 X. 24.

„ 惟 IX. 7, 40, 41; XIII. 27.

„ 尾 XI. 29.

wei	威	XI. 54, 55.
„	未	*passim.*
„	味	V. 9.
„	位	VI. 34.
„	薈	IX. 17.
„	委	VII. 6, 11; IX. 38.
„	微	VI. 9; XI. 66; XIII. 17, 18.
„	圍	III. 8; VII. 36; VIII. 2; XI. 1, 9, 14, 45, 50, 51.
„	葦	IX. 17.
wên	文	IX. 43.
„	愠	XII. 18, 20.
„	聞	XI. 18, 30.
„	聞	I. 11; II. 5; IV. 10; VII. 23; XIII. 19.
wo	我	VI. 11, 12, 13, 14, 27; X. 2, 6, 7, 8, 10; XI. 4, 5; XIII. 21.
„	臥	XI. 28.
wu	無	*passim.*
„	勿	VII. 32, 33, 34, 35, 36; IX. 4, 15; X. 9, 11; XI. 22, 57; XII. 7.
„	五	*passim.*
„	伍	III. 1.
„	吾	*passim.*
„	務	II. 15; VIII. 8.
„	侮	VIII. 12.
„	惡	IX. 11; XI. 27, 30.
„	武	IX. 40, 43.
„	吳	XI. 30.*
Ya	牙	XIII. 26.*
yai	隘	X. 1, 8; XI. 9, 45.

yang	羊	XI. 39.
„	佯	VII. 34.
„	養	II. 17; IX. 12; XI. 22, 55.
„	陽	I. 7; IX. 11, 13; X. 3, 10.
yao	要	VII. 32; XIII. 27.
yeh	也	*passim.*
„	業	VIII. 10.
„	野	XI. 21.
„	謁	XIII. 20.
„	夜	VII. 7, 26; IX. 32; XII. 11.
yen	焉	XI. 23.
„	言	VII. 23; IX. 35; XI. 57.
„	嚴	I. 9; X. 18.
„	驗	XIII. 5.
„	煙	XII. 2.
„	偃	XI. 28.
yin	引	III. 16; X. 7, 11.
„	陰	I. 7; VII. 19; IX. 11.
„	飲	IX. 30.
„	闉	III. 4.
„	殷	XIII. 26.*
„	因	I. 17; II. 9; VI. 26, 31, 33; XII. 2, 5; XIII *passim.*
ying	營	IX. 23.
„	盈	X. 8, 9.
„	嬰	X. 25.
„	應	VI. 28; XII. 5, 6.
„	迎	IX. 4, 5, 6, 16, 39.
yo	約	VI. 15; IX. 26; XI. 25.
yu	有	*passim.*
„	右	VI. 17, 20; IX. 9, 13; XI. 30; XIII. 20.

yu 由 VIII. 3; XI. 9, 19.

„ 誘 I. 20; VII. 4; IX. 28.

„ 又 III. 4; IX. 39.

„ 幽 XI. 35.

yung 用 *passim.*

„ 勇 I. 9; IV. 12; V. 17, 18; VII. 25; XI. 28, 32.

yü 雨 IX. 14.

„ 於 *passim.*

„ 予 V. 19.

„ 御 III. 17.

„ 禦 VI. 10; XI. 51.

„ 愚 XI. 36.

„ 遇 X. 17; XI. 30.

„ 虞 III. 17; XI. 19.

„ 豫 VII. 12; XI. 52.

„ 迂 VII. 3, 4, 22; XI. 9, 37.

yü 餘 IV. 6; VI. 24; XI. 27.

„ 欲 III. 17; VI. 11, 12; IX. 5, 14, 19, 38; XIII. 20.

„ 與 *passim.*

yüan 遠 I. 8, 19; II. 10; VI. 20; VII. 31; IX. 3, 16, 19; X. 1, 12, 21.

„ 原 II. 13.

„ 捐 VII. 6.

„ 圓 V. 16, 22, 23.

yüeh 曰 *passim.*

„ 月 III. 4; IV. 10; V. 6; VI. 34.

„ 越 VI. 21*; XI. 30*, 43.

„ 悅 XII. 20.

yün 紜 V. 16.

„ 輼 III. 4.

„ 運 XI. 22.

INDEX

[The numerals refer to pages]

———

Abstract ideas of degree, 50.

Accessible ground, 100, 101, 119.

Accommodating oneself to the enemy, 145, 148.

Adaptation to circumstances, 23.

Aides-de-camp, 171.

"Aids to Scouting," quoted, 88, 89, 107, 164.

Alliances, 60, 119, 140, 142.

Allotments of land, 62.

Alps, crossings of the, 57.

Amiot, Père, vii, 1.

Anger, succeeded by gladness, 159.

Army, divisions of the, 17, 33.

Army on the march, 140.

Arsenals, burning of, 151.

"Art of War," quoted by Han Hsin, 144.

Art of war in a nutshell, 44.

Athletics, 124.

Attack, skill in, 28.

Attack and defence, 25, 44.

Autumn hair, 29.

Baden-Powell, General. See "Aids to Scouting."

Baggage, 58.

Baggage-train, 60.

Baggage-trains, burning of, 151.

Bait, offered by the enemy, 68.

Balancing of chances, 31.

Banners. See Flags and banners.

Bases of supply, 60.

Beasts, startled, sign of surprise attack, 89.

Belgians at Waterloo, 130.

Benevolence to spies, 170.

Biot's Chou Li, ix.

Birds rising, sign of ambuscade, 89.

Blücher, 48.

Bluster, 95.

Boers, 18.

"Book of Army Management," 63.

Buff-coats, 58.

Burning one's boats, 133.

Calamities, six, 105.

Calthrop, Capt.: his edition of Sun Tzŭ's text, xxxii; his translation of Sun Tzŭ, viii; quoted, passim.

Camp, shifting, 133.

Camping, 80 sqq.

Cannae, battle of, 11.

Casinum, 140.

"Catalogue of Chinese Books," xxxiv.

Chan Kuo Ts'ê, quoted, 10; referred to, xxiv.

Chan Tou Ta Chia Ping Fa, xviii.

Chang Ao, a commentator, xlii.

Chang Hsiu, 69.

Chang Liang, li, 109, 116.

Chang Ni, 144.

Chang Shang-ying, lii.

Chang Shou-chieh, xvi, xvii.

Chang Tsai, li.

Chang Tzŭ-shang, a commentator, xli.

Chang Yü's commentary on Sun Tzŭ, xl; quoted, 5, 8, 9, 11, 20, 21, 22, 24, 25, 27, 30, 33, 34, 35, 39, 42, 44, 46, 49, 50, 51, 55, 56, 58, 60, 63, 64, 65, 66, 68, 69, 72, 73, 74, 75, 76, 77, 78, 80, 81, 82, 83, 85, 87, 88, 89, 90, 92, 94, 97, 99, 103, 105, 107, 109, 111, 112, 119, 124, 125, 126, 127, 131, 132, 133, 134, 136, 139, 141, 142, 143, 145, 152, 155, 156, 158, 159, 161, 163, 167, 170, 171, 172; referred to, 6, 15, 17, 31, 36, 45, 71, 86, 95, 96, 106, 147, 153, 173.

Ch'ang mountains, 128.

Ch'ang-cho, battle of, 66.

Ch'ang-shê, siege of, 154.

Chao State, army of, 28, 143; defeated by Ch'in, 166; King of, 57.

Chao Chan, 106.

Chao Kua, xlviii, 166.

Chao Shê, famous march of, 57, 136; his use of spies, 166.

Chao Yeh, xiv.

Chao Ying-ch'i, 78.

Chao Yüan-hao's rebellion, xli.

Ch'ao Kung-wu, quoted, xxxvi, xxxvii, xxxviii, xl, xli.

Chariots, 9, 91.

Chariot fighting, 15, 16.

Chariot wheels, burying of, 129.

Chavannes, M.: his "Mémoires Historiques" referred to, xiii, xvi, xlvi, 57.

Ch'ên Chên-sun, quoted, xxiii.

Ch'ên Hao's commentary on Sun Tzŭ, xxxvi, xxxviii; quoted, 30, 44, 56, 62, 65, 69, 73, 81, 93, 97, 106, 108, 110, 117, 118, 122, 124, 133, 136, 141, 147, 152, 170, 175; referred to, 18, 68.

Ch'ên-ts'ang, siege of, 94.

Chêng, principality of, 104, 116.

Chêng and ch'i. See Tactics, direct and indirect.

Chêng Ch'iao, xl.

Chêng Hou, quoted, xliii.

Chêng Hsüan's commentary on the Chou Li, xviii.

Chêng Tuan, xlii.

Chêng Yu-hsien's I Shuo, xxxii, xxxiv; referred to, 36, 53, 58, 70, 136.

Ch'êng-an, city of, captured by Han Hsin, 28.

Ch'êng-hung, battle of, 78.

Ch'êng T'ang, xvi, 173, 175.

Chi Hsieh, editor of commentaries on Sun Tzŭ, xxxviii, xli.

Chi-mo, siege of, 90.

Chi T'ien-pao's edition of Sun Tzŭ, xxxi, xxxii, xxxiii, xxxvi, xxxvii.

Ch'i State, xii, xvi, 128.

Ch'i Chieh, 90.

Chia Hsü, a commentator, xli.

Chia-ku, meeting at, xlvii.

Chia Lin's commentary on Sun Tzŭ, xxxvi, xxxviii; quoted, 20, 30, 34, 46, 50, 57, 72, 75, 76, 86, 92, 94, 95, 97, 117, 120, 133, 143, 148, 152, 157, 175; referred to, 51, 55, 62, 65, 96, 108, 164.

Chia Yü, referred to, xlvii.

Chiang-ling, town of, 111.

Chiang Yüan, a spurious work, lii.

Chieh Kuei, the tyrant, 173.

Chieh-li, a Turkish Khan, 167.

Ch'ien Ch'io Lei Shu, liii.

Ch'ien Fu Lun, referred to, xxiv.

Ch'ien Han Shu, quoted 81, 145, 167; referred to, li, 28, 34, 57, 69; bibliographical section of, quoted, xvii, xix, li; referred to, xviii, xx, liii.

Ch'ih Yu, 84.

Chin State, xii, xvi, 106.

Chin Shu, quoted, 78, 116; referred to, 123, 165.

Ch'in State, 142.

China's experience of war, xliv.

Chinese characters, elasticity of, 159.

Chinese sentiment opposed to militarism, xliv.

Ching, Duke of Ch'i, xv.

Ching-chou Fu, 123.
Ching-hsing pass, battle of, 143.
Ching K'o, 127.
Ching Wang, period of, xxiii.
Chiu T'ang Shu, referred to, 104, 167; bibliographical section of, referred to, liii.
Chou Ch'in Shih I Tzŭ, text of Sun Tzŭ in, xxxi.
Chou dynasty, 174.
Chou Hsin, the tyrant, l, 174.
Chou Li quoted, 14, 55, 60, 68, 92, 146; referred to, xxxix, xlviii, 64; Biot's translation of, ix.
Chu Chih-wu, xxi.
Chu Fu's edition of Sun Tzŭ, xvii, xxxi.
Chu Hsi, corrected by Legge, 32; quoted, xliii, xlvii.
Chu-ko Liang, 46, 51, 74, 82, 117, 122; supposititious works of, lii.
Chu-ko Wu-hou. *See* Chu-ko Liang.
Ch'u State, xii, xiii, xvi, 124; the hereditary enemy of Wu, xxvii; Viscount of, 110.
Chuan Chu, xxi, 128.
Chuan Shê-chu. *See* Chuan Chu.
Chuang, Duke of Lu, 66.
Chuang, Prince of Ch'u, 141, 162.
Chuang Tzŭ, referred to, 29, 85.
Chung Yung, xix.
Circumstances, art of studying, 68.
Classics, compared with Sun Tzŭ, xliii.
Clearness of orders, 107.
Clever fighter, the, 29, 41, 42.
Cohesion, 134.
Collapse, one of the six calamities, 105, 106.
Columns, marching in, 49.
Commander, the, 2, 3. *See also* General.
Commander-in-chief, killing the, 145; presence of mind of the, 66.
Commentary, native, on Sun Tzŭ, ix, xxxiv *sqq.*
Communications, line of, 101, 119.
Compactness, 61.

Confucius, and the art of war, xlvi, xlvii, xlviii; contemporary with Sun Tzŭ, xxx; violates extorted oath, xlix.
Constellations, 153.
Contentious ground, 115, 118, 136.
Contraction and expansion, 134.
Conventional canons of warfare, 148.
Co-operation, 129.
Council-chamber, sternness in the, 146.
Country, natural features of, 60.
Courage, one standard of, 130.
Courant's "Catalogue des Livres Chinois," lii.
Cowardice, 78.
Critical ground, 134, 135.
Cromwell's use of spies, 164.
Cross-divisions, 100.
Cunning, 145.

Danger, bracing effect of, 139, 145.
Dangerously isolated ground, 72.
Deception, war based on, 6, 132.
Decision, 37, 38.
Deductive calculation, 163.
Defence, skill in, 27.
Deliberation, 63.
Demosthenes, the Athenian general, 118.
Desertion, 134, 136.
Desperado, running amok, 125.
Desperate foe, not to be pressed, 69, 94.
Desperate ground, 72, 114, 117, 120, 125, 126, 135, 138, 143.
Deviation, artifice of, 57, 63.
Difficult grond, 71, 117, 120, 137.
Disaffection, signs of, 95.
Discipline, 2, 3, 4, 98, 111.
Disorder, seeming, 38.
Disorganisation, 105, 107.
Dispersive ground, 114, 118, 135.
Disposition of troops, 26.
Dispositions, concealment of, 51, 52; knowledge of the enemy's, 163.
Dissimulation, 61.

Dividing the enemy, 47.
Divination, to be prohibited, 126.
"Divine manipulation of the threads,"
164.
Door, left open by the enemy, 147.
.Doorkeepers, 171.
Drums, 34, 64, 65.
Dust, sign of the enemy, 89.

Earth, as opposed to Heaven, 2, 4,
27, 28, 113; six principles con-
nected with, 104.
Economy, false, 162.
Energy, 38, 39, 41; concentration
of, 124.
Entangling ground, 100, 102.
Enterprise, the spirit of, 157.
Enticing the enemy, 102.
Êrh-chu Chao, 138.
Êrh Ya, quoted, 94.
Excellence, supreme, 17; the acme
of, 28.
Expenditure on war, 9, 10, 160.

Fabius Cunctator, 11, 120.
Facile ground, 115, 118, 135, 136.
Fan Chü's use of spies, 166.
Fei River, battle of the, 25.
Fêng Hou, lii, 84.
Fêng I, a student of Sun Tzŭ, xlii.
Fire, as an aid to the attack, 156;
dropping, 151, 152; five ways of
attacking with, 150; material for,
152; proper seasons for attacking
with, 152, 153; to be started on
the windward side, 155.
Five advantages, the, 72, 74, 75.
Five cardinal tastes, 36.
Five cardinal virtues, 3.
Five classes of State ceremonial, xlviii.
Five dangerous faults, 77.
Five developments in attacking with
fire, 153 sqq.
Five elements, the, 53.
Five essentials for victory, 23, 24.
Five factors in war, 1.
Five musical notes, 36.

Five Pa Wang, xlix, 141.
Five primary colours, 36.
Flags and banners, 16, 34, 64, 65.
Flat country, campaigning in, 83, 84.
Flight, 105.
Foraging, 12, 15, 123, 161.
Foreknowledge, 163.
Forestalling the enemy, 147.
Forethought, want of, 97.
"Forty-one Years in India," referred
to, 35.
Four seasons, the, 54.
Frederick the Great, quoted, 48,
168, 169.
Frontier passes, 146.
Frontal attacks, 45.
Fu Ch'ai, xvi.
Fu Chien, 25, 115.
Fu-ch'u, King of Ch'u, 124.
Fu Kai, xxiii, xxix.
Fu-k'ang-an, 63.
Fu Yen-ch'ing, 69, 70.

General, the, 4, 5, 7, 8, 15, 16, 19,
21, 44, 55, 66, 77, 98, 107, 109,
110, 130, 131, 134, 157, 159, 163,
171, 174.
Generals, professional, xxii.
Generalship, degrees of, 17, 18; the
highest, 48.
Giles' Biographical Dictionary, quo-
ted, 128.
Giles' Chinese-English Dictionary,
referred to, 57, 134.
Gongs, 34, 64.
Grant, General, 47.
Great Wall of China, xliv.
Greeks, Homeric, 9.
Grindstone and egg, 35.
Ground, high and low, 84; of inter-
secting highways, 71, 116, 119,
135, 137; proper use of, 130.
Grounds, the nine, 114, 134, 138.
Guides, local, 60, 140.

Han, red banners of, 144.
Han Chih. See Ch'ien Han Shu,
bibliographical section of.

Han Kuan Chieh Ku, quoted, xx.

Han Hsin, xliv, 28, 33, 34, 81, 143, 167; a student of Sun Tzŭ, xlii; quoted, 68.

Han Shu. See Ch'ien Han Shu.

Hannibal, 11, 57, 66, 120, 140.

Hasty temper, 78.

Hearing, quick, 29.

Heaven, 2, 4, 28, 113.

Heights, precipitous, 100, 103.

Hemmed-in ground, 72, 117, 120, 135, 137.

Henderson, Col., quoted, 6, 42, 48, 52, 59, 101, 130, 131.

Herodotus, referred to, 129.

Ho Ch'ü-fei, xl.

Ho Kuan Tzŭ, referred to, xxiv.

Ho Lu (or Ho Lü), King of Wu, xi, xiii, xvi, xvii, xviii, xxvi, 5, 128.

Ho Shih. *See* Ho Yen-hsi.

Ho-yang, night ride to, 65.

Ho Yen-hsi's commentary on Sun Tzŭ, xl; quoted, 11, 14, 16, 18, 21, 29, 30, 34, 56, 69, 74, 110, 115, 116, 122, 147, 165, 166, 167, 168, 174; referred to, xvii, 31, 43, 62, 152.

Horses, tethering of, 129.

Hou Han Shu, quoted, 10, 94, 132, 139, 151, 155; referred to, xlii.

Hsi, the graduate, xxxiii.

Hsia dynasty, 174.

Hsiang, Duke of Sung, xlix, 141.

Hsiang Chi, xlix, 133.

Hsiang Liang, xlix.

Hsiang Yü. *See* Hsiang Chi.

Hsiao State, 110.

Hsiao Chi, a commentator, xli.

Hsiao Hsien, 123.

Hsiao I, 153, 166.

Hsiao Shih-hsien. *See* Hsiao I.

Hsieh An, 25.

Hsieh Yüan, a commentator, xlii.

Hsien Hu, 106.

Hsin-ch'êng, town of, 122.

Hsin Hsü, xiv.

Hsin Shu (by Ts'ao Kung), xix, xxxvi.

Hsin Shu (a work attributed to Chu-ko Liang), lii.

Hsin T'ang Shu, referred to, 65, 104, 105, 123, 167; bibliographical section of, referred to, xviii, liii.

Hsing Li Hui Yao, quoted, xliii, xlviii.

Hsing Shih Pien Chêng Shu, xv.

Hsiung-nu, 39, 139, 150.

Hsü Ch'ieh, quoted, 160.

Hsü-chou, invaded by Ts'ao Ts'ao, 73.

Hsü Wên Hsien T'ung K'ao, liii.

Hsüan Tsung, T'ang Emperor, xxxii.

Hsün Tzŭ, quoted, 80.

Hsün Ying, 73.

Hu Yen, xiii.

Hua-pi, city of, 73.

Hua-yin temple, xxxii.

Huai-nan Tzŭ, plagiary of Sun Tzŭ, xxiv; quoted, xiv.

Huan, Duke of Ch'i, 128, 141.

Huan Ch'ung, 25.

Huan Hsüan, 78.

Huang Ch'ao Ching Shih Wên Pien, liii.

Huang Chih-chêng, a commentator, xlii.

Huang Jun-yü, a commentator, xli.

Huang Mei, 78.

Huang-shih Kung, li; quoted, 109, 126.

Huang Ti. *See* Yellow Emperor.

Huang-fu Sung, 94, 154, 155.

Human nature, to be studied, 134.

Humanity, misplaced, xlix; soldiers to be treated with, 98.

Husbanding one's strength, 67.

Husbandry, impeded by war, 161.

I river, 127.

I Chih, 173, 174, 175.

I Ching, quoted, xv.

I-chou, 165.

I-ho, 115.

I Pu Chê Chung, xliii.

I Shuo. See Chêng Yu-hsien.

I-wu pass, 115.

I Yin. *See* I Chih.

Iliad, heroes of the, 127.

Impoverishment of the people, 13, 14.

Induction from experience, 163.

Inhumanity, the height of, 162.

Insubordination, 105.

Intuition necessary in dealing with spies, 169.

Invading force, principles to be observed by an, 123.

Jackson, Stonewall, 59, 131.

Jan Yu, disciple of Confucius, xlvi, xlviii.

Jang, siege of, 69.

Jingles, 149, 158.

Jn-nan, 111.

Julius Caesar, 12; his "De Bello Gallico" referred to, 108.

Junction of forces, 48.

K'ang Hsi's dictionary, referred to, 10, 18, 35, 68, 95, 117, 152, 157, 160.

Kao-ch'ang, 115.

Kao-fan. *See* Hu Yen.

Kao Huan, afterwards Emperor, 137.

Kao Kêng, 151.

Kao Tsu, first Han Emperor, 33, 39, 119.

Kao Tsu, Sui Emperor, 168.

Kao-wu pass, 115.

Khitans, 69.

Khotan, 132.

Kiangnan rebels under Sui dynasty, 151.

Kindness to the soldiers, 110, 111.

Kou Chien, King of Yüeh, xvi, 50

Ku Chin T'u Shu Chi Ch'êng, quoted, xvi, xxxvii, xxxix; referred to, xix, xli, li, liii. *See also* Sun Tzŭ, *T'u Shu* text of.

Kuan Chung, 128.

Kuan Tzŭ, xxi.

Kuang, King of Shan-shan, 139, 151.

Kuang Po Wu Chih, liii.

Kuang Wu, Han Emperor, li.

Kuei-ku Tzŭ, li.

K'uei-chou, 123.

K'un Wai Ch'un Ch'iu, xxxvi.

Kung-sun Hung, lii.

Kuo Ch'ao Shih Jên Chêng Lüeh, xxxii.

Kuo Hsün, 151.

Kutcha, King of, 132.

Ladder, kicking away the, 133.

Ladysmith, relief of, 79.

Land-tenure, ancient system of, xxv, 161.

Lao Tzŭ, the Tao of, 2; quoted, 155, 158. See also *Tao Tê Ching*.

Legge's "Chinese Classics," referred to, ix, xxiv, 23, 32.

Lengthy operations, 10, 11.

Li, length of the, 9.

Li Chi, referred to, 23, 147.

Li Ching, the general, xliv, 41, 123, 167; quoted, 35, 66, 87, 111, 118; supposed author of a work on war, lii.

Li Ching Ping Fa, lii.

Li Chu, 29.

Li Ch'üan's commentary on Sun Tzŭ, xxxvi; quoted, 9, 11, 18, 21, 22, 24, 25, 28, 30, 32, 34, 38, 46, 49, 50, 51, 55, 60, 65, 67, 68, 72, 73, 81, 83, 84, 89, 92, 97, 105, 106, 110, 113, 114, 115, 117, 118, 119, 136, 142, 150, 158, 163, 167; referred to, 52, 95, 123, 127, 151.

Li Hsiang, 165.

Li Hsiung, 165.

Li I-chi, 167.

Li Kuang-pi, 65.

Li Ling, 154.

Li Shih-min, afterwards the Emperor T'ai Tsung, xliv, lii, 35, 104, 167.

Li Shou-chêng, 70.

Li Tai Chi Shih Nien Piao, quoted, 70, 116, 166.

Li T'ê, 165.

Li Ts'ai, a commentator, xlii.

Li Wei-kung. *See* Li Ching.

Li Wei Kung Wên Tui, lii.
Liang, kingdom of, 94.
Liang-chou, 115.
Liang Hsi, 115.
Lien P'o, 57, 166.
Lin-chin, in Shensi, 34.
Lin Hsiang-ju, 166.
Line of least resistance, 53.
Liu Chou-tzŭ, 53.
Liu Hsiang, quoted, xiv, xxiv.
Liu Pei, 59.
Liu Piao, 69.
Liu T'ao (attributed to T'ai Kung), xxi, 1, li, 144, 174; quoted, 22, 62, 78, 84.
Liu Yü, 78.
Livy, quoted, 66, 120, 140.
Lo Shang, 165.
Lo-yang, 104.
Logs and stones, rolling, 41.
Longevity, 127.
Lou Ching, 39.
Lu State, 128.
Lu Tê-ming, quoted, li.
Lü Kuang, 115.
Lü Mêng, a disciplinarian, 111; a student of Sun Tzŭ, xlii.
Lü Pu, xxxv.
Lü Shang, known as T'ai Kung, 1, 174, 175. See also *Liu T'ao*.
Lü Shih Ch'un Ch'iu, referred to, xxiv, 37.
Lü Wang (or Lü Ya). See Lü Shang.
Luan Yen, 106.
Lun Yü, quoted, xv, 146; referred to, xlvii, xlix, 47, 64, 156.
Lung Chü, 81.

Ma Lung, lii.
Ma Tuan-lin, xl. See also *Wên Hsien T'ung K'ao*.
Ma Yüan, 80.
Maiden, coyness of a, 148.
Mansfield, Lord, 143.
Mantlets, 14, 18.
Marches, forced, 59.
Marengo, battle of, 57.

"Marshal Turenne," quoted, 73, 169; referred to, 61.
Marshes, 60.
Measures, of land, 14; of length, 32; of weight, 15, 32.
Mei Yao-ch'ên's commentary on Sun Tzŭ, xxxviii; quoted, 4, 6, 7, 11, 29, 34, 38, 40, 44, 47, 61, 63, 79, 84, 85, 86, 93, 94, 95, 96, 100, 102, 121, 129, 130, 131, 135, 136, 137, 138, 141, 145, 147, 148, 153, 155, 157, 161, 162, 163, 164, 168, 169, 170, 174; referred to, 15, 23, 43, 46, 51, 106, 151.
"Mémoires concernant les Chinois," quoted, vii.
"Mémoires Historiques," referred to, xvi. *See also* Chavannes.
Mencius, quoted, xxv, xliii, 14, 85; referred to, 29, 32, 112, 148.
Mêng K'ang, xxxvi.
Mêng Shih's commentary on Sun Tzŭ, xxxvi; quoted, 2, 11, 15, 61, 77, 78, 116, 137, 147.
Mêng Ta, 122.
Method, 2, 3, 31.
"Military Classic," 144.
Military tactics like water, 53.
Military virtues, 22.
Misfortune, three ways in which a ruler can cause, 21 *sqq*.
Mistakes, making no, 30.
Modern text of Sun Tzŭ. *See* Sun Tzŭ.
Modification of plans, 5.
Moltke, 17.
Moods, art of studying, 67.
Moral Law, the, 2, 4, 31.
Mounds, used in sieges, 19.
Mountains, 80.
Movable shelters, 18.
Mu, Duke of Ch'in, 141.
Mu-so, an instrument of torture, xlvi.
Mu T'ien Tzŭ Chuan, 152.
Mystification of one's men, 131.

Nang Wa, xiii.

Napoleon Bonaparte, 5, 12, 148; his passage across Alps, 57; not hampered by central authority, 24; his "Maximes de Guerre," quoted, 84, 109; his "Pensées," quoted, 101.

Nelson, at Trafalgar, 37.

Nervousness, a sign of, 93.

Nicias, the Athenian general, 118; speech of, quoted, 125.

Night-fighting, 65.

Nine grounds (or situations), the, 72, 114.

Nine punitive measures, the, xxxix.

Nine variations, the, 71, 72, 74, 138.

"North hill", battle of the, 57.

O-yü, town of, 57.

Omens, not to be regarded, 126.

Onset of troops, 37, 38.

Open ground, 116, 119, 137.

Opportunism, xlix.

Orders, not to be divulged, 142, 143.

Original text of Sun Tzŭ. *See* Sun Tzŭ.

Ou-yang Hsiu, quoted, xxxiv, xxxv, xxxviii.

Overawing the enemy, 141.

Over-caution, 158.

Over-solicitude for one's men, 79.

Pa Chên T'u, xviii.

Pa Wang, the five, 141.

Pan Ch'ao, 63; at Shan-shan, 139, 150; his attack on Yarkand, 132, 167.

P'an Kêng, 173.

P'ang Chüan, xii, 40.

Passes, narrow, 100, 103.

Peace, the true object of war, 162.

Pei Ch'i Shu, referred to, 138.

Pei Lun, xl.

Pei T'ang Shu Ch'ao, 25, 36, 64, 67.

P'ei Hsing-chien, 103.

P'ei Wên Yün Fu, quoted, 94; referred to, xlvi, 69, 146.

Pelliot, M., xxxvi.

Pi, battle of, 106.

Pi I-hsün, xviii, xxvi, xxxiv. See also *Sun Tzŭ Hsü Lu*.

Pi Kua, xxxiii.

Pi-yang, city of, 73.

P'i, siege of, 165.

Picked soldiers in front rank, 107, 108.

Ping Fa Tsa Chan, xviii.

Ping Shu Yao Chüeh, 67.

Pique, battles not to be fought out of, 158.

Pitfalls, 60.

Plagiaries of Sun Tzŭ, xxiii, xxiv.

Plans, baulking the enemy's, 17; change of, 5, 132.

Plataea, battle of, 129.

Playfair's "Cities and Towns of China", referred to, 57.

Plunder, 62.

Po Ch'i, xliv, 117, 166.

Po Chiang Chuan, xli.

Po P'ei, xiii, xxiii, xxix.

Po-têng, battle of, 39.

Po-ts'ai, a leader of the Yellow Turban rebels, 154.

Po Ya, referred to, 160.

P'o-t'ai, a spy, 165.

Polybius, referred to, 120.

Port Arthur, siege of, 19.

Presence of mind, 66.

Punishment, 95, 97, 98.

Rabbits, not indigenous to China, 149.

Rapidity, 12, 61; the essence of war, 122.

Rewards, 15, 95, 142.

Reward and punishment, constancy in, 4.

Riches, soldiers not to acquire, 127.

River, crossing a, 129.

River warfare, 81, 82.

Roberts, Lord, night march of, 35; on Sun Tzŭ, xlii.

Rout, 105, 107.

Ruin, one of the six calamities, 105, 106.

Ruler, military commander independent of the, 109; the enlightened, 157, 159, 174.

Rules of warfare, conventional, 148.

Salt-marshes, 83.

San Kuo Chih, quoted, 69, 111; referred to, xxxv, xli, xlii. See also *Wei Chih*.

San Lüeh, li; quoted, 62, 158.

San Shih Êrh Lei Ching, xviii.

San Ts'ai T'u Hui, liii.

San-yüan, 79.

"Science of War," quoted, 101, 130.

Scouts, 88, 89.

Screens, grass, 88.

Secrecy, 45, 131.

Secrets, divulged by a spy, 170.

Sedan, capitulation of, 17.

Self-possession, 67.

Sensitiveness in a general, 79.

Sentries, 171.

Serious ground, 117, 119, 135, 137.

Seven considerations, 1, 4.

Sha-yüan, 168.

Shan-shan, 139; King of, 150, 151.

Shang dynasty, 173.

Shên, Duke of, 110.

Shên-wu of Ch'i, 168.

Shên Yu, a commentator, xli.

Shepherd driving sheep, 133.

Sheridan, General, 47.

Shih Chi, objection to the chronology of, xxvi; quoted, xi, xiii, xv, xx, xlv, 40, 58, 80, 84, 90, 124, 128; referred to, xvi, xxii, xxiv, xxxiv, xlvi, xlvii, xlix, l. *See also* Ssŭ-ma Ch'ien.

Shih Ching, quoted, xvi, 61, 62; referred to, 14.

Shih Huang Ti, 127, 142.

Shih K'uang, 29.

Shih Liu Ts'ê, lii.

Shih Ssŭ-ming, the rebel leader, 65.

Shu Ching, quoted, xv; referred to, xlvii, xlviii.

Shu Lu Chieh T'i, xxiii.

Shuai-jan, the, xxvi, 128, 129.

Shuo Wén, quoted, 94, 117, 160.

Sicilian expedition, 118.

Sieges, 10, 18, 19, 73.

Sight, sharp, 29.

Signal-fires, 65.

Signals, 33.

Signs, observation of, 88.

Situations, the nine. *See* Nine grounds.

Six Chancellors of the Ch'in State, 142.

"Six States" period, xxii.

Skilful fighter, the, 30.

Skilful leaders of old, 120.

Solidarity of troops, 123.

Sôphanes at Plataea, 129.

Sovereign, the, 55; the wise, 163.

Spies, xlix, 52, 147, 148; converted, 90, 166, 172, 173; doomed, 167, 172, 173; five classes of, 164; Frederick's classification of, 168; importance of, 175; intimate relations to be maintained with, 168; inward, 165, 172; local, 164, 172; surviving, 167, 172; to be properly paid, 162, 169.

Spirit, an army's, 65, 66.

Spirits, 163.

"Spy," evolution of the character meaning, 160.

Spying, end and aim of, 173.

Ssŭ K'u Ch'üan Shu Chien Ming Mu Lu, quoted, l, li, lii.

Ssŭ K'u Ch'üan Shu Tsung Mu T'i Yao, quoted, xx, xli, l; referred to, xl, lii, liii.

Ssŭ-ma Ch'ien, xiv, xx; quoted, xi, xii, xlv; credibility of his narrative, xxvi; his letter to Jên An, referred to, xlvi; his mention of the 13 chapters, xxx. See also *Shih Chi*.

Ssŭ-ma Fa, l; quoted, xvi, 14, 17, 78, 126, 143.

Ssŭ-ma I, 46, 51, 122.

Ssŭ-ma Jang-chü, xxii, l, 98.

Stagnation, 157.

Standard text of Sun Tzŭ. *See* Sun Tzŭ.

Stellar Mansions, the twenty-eight, 153.

Stonewall Jackson, biography of, quoted, 42, 59, 131.

Strategy and tactics, 52.

Strength, great, 29.

Stupidity, to be feigned, 145.

Su Hsün, quoted, xlii.

Su Shu, an ethical treatise, li.

Subdivisions of an army, 17, 33, 39.

Sui Shu, quoted, 151; bibliographical section of, quoted, xviii, xli; referred to, xxxvi, liii.

Sun Hao, a commentator, xli.

Sun Hsing-yen, xxxii; his edition of Sun Tzŭ, ix; his preface, xxxiv; quoted, xvi, xxix, xxx, xxxi, xxxii, xxxiii, xxxvi, xlviii.

Sun Pin, xii, xv, xvi, 40.

Sun Tzŭ, archaic words in, xxiv; bibliographical description of edition used, xxxiv; corruptions in the text of, xxxi; difficult passages in, xxxiv; state of the text, 138; probable date of the work, xxviii.

— Modern text, 25, 26, 27, 33.

— Original text, xxxii, xxxiii, 2, 16, 27, 29, 43, 47, 53, 58, 62, 64, 67, 84, 86, 87, 88, 91, 92, 95, 98, 113, 119, 121, 153, 154, 168.

— Standard text, xxxiv, 10, 58, 91, 95, 117, 127, 164.

— *T'ai I Tun Chia* text, xxxvi.

— *T'u Shu* text, xxxi, 16, 21, 25, 29, 30, 32, 33, 35, 37, 40, 43, 46, 47, 50, 52, 58, 64, 67, 69, 84, 87, 91, 92, 94, 95, 96, 105, 110, 114, 117, 120, 121, 133, 135, 140, 145, 146, 153, 159, 164, 167, 168, 171, 172, 175.

— *T'ung Tien* text, xxxiii, 1, 10, 12, 19, 22, 23, 25, 41, 45, 47, 50, 53, 58, 59, 62, 64, 65, 67, 68, 74, 77, 81, 83, 85, 86, 87, 88, 89, 91, 92, 93, 94, 95, 98, 101, 104, 108, 112, 113, 117, 119, 136, 137, 152, 153, 158, 159, 164, 167, 170, 171, 172.

— *Yü Lan* text, xxxiii, 3, 7, 10, 12, 14, 15, 19, 25, 27, 37, 42, 45, 47, 50, 52, 53, 62, 64, 67, 68, 77, 81, 83, 84, 85, 86, 87, 88, 89, 92, 93, 94, 95, 98, 108, 112, 121, 129, 141, 153, 158, 159, 161, 164, 167, 170, 171, 172.

Sun Tzŭ Hsü Lu, xviii, xxxiv; quoted, xxiii, xxiv, 118.

Sun Tzŭ Hui Chêng, xlii.

Sun Tzŭ Ts'an T'ung, xlii.

Sun Tzŭ Wên Ta, xvii.

Sun Wu, a practical soldier, xxv; conjectural outline of his life, xxix; not a man of eminent position, xxviii; probable origin of the legend connected with, xxix; Ssŭ-ma Ch'ien's biography of, xi; supposititious works of, xvii, xviii. See also *Sun Tzŭ*.

Sun Wu Sun Tzŭ, xvii.

Sung Shih, referred to, xlii; bibliographical section of, xvii, xxxi, xxxvi, lii, liii.

Superstitious doubts, 126.

Supplies, 137, 161; line of, 101.

Ta-hsi Wu, 168.

Ta Ming I T'ung Chih, quoted, xxxii.

Taboo character, 124.

Tactical manœuvring, 56.

Tactician, the skilful, 128.

Tactics, direct and indirect, 20, 34 *sqq.;* modification of, 52, 53; not to be repeated, 52; variation of, 26, 71, 74.

T'ai Kung. *See* Lü Shang.

T'ai Kung Ping Fa, li.

T'ai P'ing Yü Lan, xvi, xxxiii, liii. See also Sun Tzŭ, *Yü Lan* text.

T'ai-po Shan-jên, quoted, 132.

T'ai Po Yin Ching, xxxvi.

T'ai Tsung, the Emperor. *See* Li Shih-min.

T'ai Yüan Ching, referred to, xxiv.

Tallies, official, 146.

T'ang, prince of, xiii.

T'ang, the Completer. *See* Ch'êng T'ang.

T'ang Chien, 167.

T'ang Shu, bibliographical section of, referred to, xxxviii, xli. See also *Hsin T'ang Shu* and *Chiu T'ang Shu*.

Tao Tê Ching, quoted, xlix, 147, 155, 158, 161.

Temple, used for deliberations, 7, 8.

Temporising ground, 100, 102.

Tenacity, 125.

Têng Ch'iang, 78.

Têng Ming-shih, quoted, xv.

Terrain, natural advantages of, 108; six kinds of, 100.

Textual criticism and emendations, 1, 7, 13, 14, 25, 29, 30, 36, 41, 43, 46, 47, 49, 71, 74, 86, 87, 91, 94, 99, 113, 117, 121, 124, 127, 133, 158, 167.

Thermopylae, 115.

Three ancient dynasties, the, xxxix.

Thucydides, quoted, 125; referred to, 118.

Ti river, 144.

T'ien Chi, 40.

T'ien-i-ko catalogue, quoted, xxxvi, xl.

T'ien Pao, xv.

T'ien Pu, 105.

T'ien Tan, defender of Chi-mo, 90, 120, 155; his use of spies, 166.

Time, value of, 12; waste of, 157.

Tou Chien-tê, King of Hsia, 104.

Tou Ku, 151.

Trafalgar, battle of, 37.

Training of officers and men, 4.

Trebia, battle of the, 66.

Ts'ai, prince of, xiii.

Ts'ao Kuei, mentioned in the *Tso Chuan*, xxi; on the advantage of spirit, 66; threatens Huan Kung, 128.

Ts'ao Kung or Ts'ao Ts'ao, xix, xxxi, xxxvi, xlii, xliv, 4, 59, 69, 76, 151;

his commentary on Sun Tzŭ, xxxv, xxxvii, xxxviii, xl; quoted, 1, 7, 9, 11, 13, 15, 17, 18, 20, 22, 24, 26, 28, 34, 35, 39, 40, 41, 44, 46, 51, 52, 55, 56, 59, 60, 67, 71, 73, 75, 76, 77, 78, 81, 84, 86, 88, 91, 94, 95, 96, 97, 98, 103, 104, 106, 111, 115, 116, 118, 119, 120, 122, 125, 126, 127, 131, 137, 140, 142, 143, 145, 146, 147, 148, 152, 154, 156, 157; referred to, 19, 43, 62, 136; his preface, xx, xxxiv; translated, xv *sqq*.

Tsêng Shên, xxiv.

Tso Chuan, delivered to Wu Ch'i, xxiv; has no mention of Sun Tzŭ, xx, xxvi, xxviii; quoted, xxvii, xxix, xlix, 19, 59, 65, 89, 97, 106, 111, 162; referred to, xxi, xlvii.

Tso Tsung-t'ang, 63.

Tsui-li, battle of, xxx.

Tu Chung-wei, 69, 70.

Tu Mu's commentary on Sun Tzŭ, xxxvi, xxxvii, xxxviii; quoted, 4, 11, 14, 15, 18, 19, 23, 26, 28, 29, 30, 31, 33, 34, 37, 39, 40, 41, 42, 44, 45, 46, 50, 52, 55, 56, 57, 59, 60, 61, 62, 64, 67, 68, 69, 75, 76, 77, 78, 80, 81, 82, 83, 84, 86, 88, 89, 90, 92, 93, 94, 95, 96, 98, 101, 105, 106, 107, 110, 111, 112, 114, 115, 118, 119, 122, 124, 126, 131, 133, 136, 137, 138, 146, 148, 149, 151, 152, 153, 154, 155, 156, 157, 158, 161, 163, 164, 165, 167, 168, 169, 171, 175; referred to, 20, 65, 73, 150; his preface, quoted, xix, xxxvii, xxxviii, xlv.

Tu Shu Chih, lii.

Tu Yu, xxxiii; his notes on Sun Tzŭ in the *T'ung Tien*, xxxvii; quoted, 4, 6, 11, 19, 23, 24, 36, 38, 47, 56, 60, 61, 62, 77, 83, 88, 91, 92, 93, 94, 95, 100, 101, 102, 103, 104, 116, 117, 120, 137, 138, 152, 153, 166, 167, 169, 171, 172; referred to, 28, 51, 74, 155, 173.

T'u Shu encyclopaedia. See *Ku Chin T'u Shu Chi Ch'êng.*
— Text of Sun Tzŭ in the. See *Sun Tzŭ.*
Tung Cho, xxxv, 94.
Tung Chou Lieh Kuo, quoted, 56.
T'ung Chih, referred to, xxxii, xxxvi, xl, xli, liii.
T'ung Tien, xvii, xxxiii, xxxvii, lii, liii. *See also* Tu Yu.
— Text of Sun Tzŭ in the. See *Sun Tzŭ.*
Turenne, Marshal, on deceiving the enemy, 61; on sieges, 73; on spies, 169.
Tzŭ-ch'an, saying of, xlix.
Tzŭ-ch'ang. *See* Nang Wa.

"Unterricht des Königs von Preussen," quoted, 168, 169.
Uxbridge, Lord, 5.

Valleys, 80.
Victory, halfway towards, 111, 112; without fighting, 17.
Virtues, the five cardinal, 3.

Wan, town of, 122.
Wang Chien, 124.
Wang Hsi's commentary on Sun Tzŭ, xl; quoted, 1, 2, 11, 13, 14, 23, 26, 33, 34, 38, 44, 52, 53, 55, 60. 61, 63, 71, 78, 84, 92, 94, 95, 96, 106, 114, 117, 119, 124, 132, 133, 135, 137, 142, 155, 157, 169; referred to, 67, 76.
Wang Kuo, the rebel, 94.
Wang Liao, 128.
Wang Ling, a commentator, xxxvii, xli. *See also* Wang Tzŭ.
Wang Shih-ch'ung, 104.
Wang T'ing-ts'ou, 105.
Wang Tzŭ, quoted, 4, 6, 24.
Wang-tzŭ Ch'êng-fu, xiii.
War, want of fixity in, 54.
Warlike prince, 141, 158.
Water, an aid to the attack, 156.

Waterloo, battle of, 5, 48, 130.
Weapons, 14.
Weeping, 127.
Wei, kingdom of, xxxv; province of, 105.
Wei river, 81.
Wei Chih (in the *San Kuo Chih*), xix, xxxvi.
Wei I, 106.
Wei Liao Tzŭ, li; quoted, 35, 73, 97, 99, 107, 125; referred to, xxiv.
Wei Po, 165.
Wei Wu Ti. *See* Ts'ao Kung.
Well-being of one's men, to be studied, 123.
Wellington, his description of his army at Waterloo, 130; on the eve of Waterloo, 5; saying, of, 110; skilful in dissimulation, 6.
Wên, Duke of Chin, 141.
Wên Hsien T'ung K'ao, quoted, xxxvii, xxxviii, xl, xli; referred to, xxi, xxiii, xxxvi, liii.
Wên-su, King of, 132.
Wên Ti, Emperor of Sui dynasty, 151.
Wên Wang, l, 174.
Western Sacred Mountain, xxxii.
Wind, days of, 153; duration of, 155.
"Words on Wellington," quoted, 5.
Wu, city of, xiv; king of, 118. *See also* Ho Lu.
Wu State, xxv, 49, 50, 129, 159; dates in the history of, xxvii, xxviii; first mentioned in history, xxvii.
Wu Ch'i, l, 64, 65, 110; compared with Sun Wu, xliii; plagiary of Sun Tzŭ, xxiv. See also *Wu Tzŭ.*
Wu Ch'i Ching, lii.
Wu Huo, 29.
Wu Jên-chi, xxxiii.
Wu-lao, heights of, 104.
Wu Nien-hu, xxxiii.
Wu-tu, town of, 165.
Wu-tu Ch'iang, 80.
Wu Tzŭ, xix, l; quoted, 24, 56, 66, 77, 80, 81, 98, 107, 115, 131, 142, 156; referred to, xxiv.

Wu Tzŭ-hsü, xxix, xlviii. *See also* Wu Yüan.

Wu Wang, xvi, 20, 175.

Wu Yüan, xiii, xxiii, 56; a spurious treatise fathered on, xxix.

Wu Yüeh Ch'un Ch'iu, quoted, xiv, xviii.

Wylie's "Notes," referred to, xli, lii.

Ya, King of Chao, 144.

Yang Han, 115.

Yang-p'ing, city of, 46.

Yangtsze river, 123.

Yao Hsiang, 78.

Yarkand, battle of, 132.

Yeh Shih or Yeh Shui-hsin, his theory about Sun Tzŭ, xxi, xxiii, xxv; on Sun Tzŭ's style, xxiv.

Yellow Emperor, the, xvi, 84.

Yellow Turban rebels, 154.

Yen, King of Hsü, xvi, xlix.

Yen Shih-ku, 167.

Yen Ti, 84.

Yen Tzŭ, quoted, 98.

Yin and *Yang,* 2.

Yin dynasty, 173, 174.

Yin Fu Ching, xxxvi, 111.

Ying, capital of Ch'u, xii, xiii, xvi, xxix.

Ying K'ao-shu, xxi.

Yo Fei, a student of Sun Tzŭ, xlii.

Yo I, 117.

Yü Hai, quoted, xlii; referred to, xxxvi, xl, lii, liii.

Yü Lan encyclopaedia. See *T'ai P'ing Yü Lan.*

— Text of Sun Tzŭ in the. See *Sun Tzŭ.*

Yüan, the two, opponents of Ts'ao Ts'ao, xxxv.

Yüan Chien Lei Han, liii.

Yüan Shao, 151.

Yüeh State, 129; compared with Wu, xxvi, 49, 50; first mentioned in history, xxvii.

Yüeh Chüeh Shu, quoted, xiv.

Yüeh Yü, xxi.

Yung Lo Ta Tien, lii.

CORRIGENDA

P. ix, *note:* For "edition" read "translation."

„ 14, line 3: For "by" read "in the."

„ 16, line 5: For "T." read "*T'u Shu.*"

„ „ § 19, *note:* Before "War" insert "Soldiers are not to be used as playthings."

„ 17, § 1: 全軍, etc. The more I think about it, the more I prefer the rendering suggested on p. 159, § 22, *note.*

„ „ § 1 *note,* and p. 78, line 6: Insert "the" before "Ssŭ-ma Fa."

„ 33, note on heading: Cf. X. § 12, where 勢 is translated "strength," though it might also be "conditions." The three words 執, 執 and 勢 have been much confused. It appears from the *Shuo Wên* that the last character is post-classical, so that Sun Tzŭ must have used either 執 or 執 in all senses.

„ 45, line 1: For "sublety" read "subtlety."

„ 63, line 4: M. Chavannes writes in the *T'oung Pao,* 1906, p. 210: "Le général Pan Tch'ao n'a jamais porté les armes chinoises jusque sur les bords de la mer Caspienne." I hasten to correct my statement on this authority.

„ 80, 9th line from the bottom: For ▯ read ▯.

„ 109, § 23, *note,* and p. 126, 5th line from bottom: For "Huang Shih-kung" read "Huang-shih Kung."

„ 124, line 7: For "Ch'ên" read "Ch'ên Hao."

„ 136, 11th line from bottom: Insert "to" before "select."

„ 152, § 2: Substitute semi-colon for full stop after "available."